"For a brief time before 1982, Dome was a corporate Camelot. Whether it was influencing politicians, sweet-talking financiers, or exaggerating its accomplishments, it had an energy, a vision, and a single-mindedness that was unmatched among companies in Canada.... Then, in this intoxicating atmosphere of success, heady with past triumphs, its officers began plotting the audacious assault to capture the Hudson's Bay Oil and Gas Company, in what was to be the biggest takeover the world had yet seen..."

—from the Introduction
by Jim Lyon,
Editor and journalist
of the *Financial Post*

D1052333

DOME

THE RISE AND FALL OF THE HOUSE THAT JACK BUILT

JIM LYON

AVON BOOKS OF CANADA
PUBLISHERS OF BARD, CAMELOT, DISCUS AND FLARE BOOKS

AVON BOOKS OF CANADA
A division of
The Hearst Corporation
2061 McCowan Road, Suite 210
Scarborough, Ontario M1S 3Y6

Copyright © 1983 by Jim Lyon
Published by arrangement with Macmillan of Canada
Library of Congress Catalog Card Number: C83-098596-4
ISBN: 0-380-68486-1

Canadian Cataloguing in Publication Data

Lyon, Jim, 1935-
    Dome: the rise and fall of the house that Jack built

Bibliography.
Includes index.
ISBN 0-7715-9777-0

1. Dome Petroleum Limited–History.   I. Title.

HD9574.C24D65   338.7′6223382′0971   C83-098596-4

First Avon Printing, June, 1984

Printed in Canada.

UNV   10   9   8   7   6   5   4   3   2   1

CONTENTS

Introduction

Part I   ON TOP OF THE WORLD/1
   Chapter 1   Operation Swampy/3
   Chapter 2   A Pyrrhic Victory/13

Part II   THE RISE OF DOME/27
   Chapter 3   Jack Gallagher: "A Class Act"/29
   Chapter 4   The House That Jack Built/40
   Chapter 5   Ottawa's Blue-eyed Boy/54
   Chapter 6   The New Breed/69
   Chapter 7   Bill Richards: Boardroom
                   Gunslinger/81

Part III   DOME AND THE NORTH/95
   Chapter 8   Out in the Cold/97
   Chapter 9   Beaufort: Fact or Fantasy?/114
   Chapter 10 Ice: The Fearful Enemy/123
   Chapter 11 A Private Armada/137
   Chapter 12 Tuktoyaktuk: The Two Solitudes/155
   Chapter 13 Frontier or Homeland/167

Part IV   A NATIONAL CRISIS/181
   Chapter 14 The House of Cards/183
   Chapter 15 Inquest: What Went Wrong?/207
   Chapter 16 Dome Looks Ahead/222

Afterword/231

Bibliography/239

Index/243

This book is dedicated
to my wife, Ingrid.

DOME

INTRODUCTION

This is not the book I intended to write about Dome Petroleum. I set out to attend a celebration but ended up at a wake. The idea for a book about Dome was born in June 1977 as I was flying to Calgary from Dome's northern base at Tuktoyaktuk on the upper rim of the North American continent. As the lakes and barren rocks of the Northwest Territories, thousands of feet below, were succeeded by the farmland of northern Alberta, I realized that there was a fine story to be written about the search for oil in the Canadian Arctic and the impact that an industrial colossus would have on the native peoples who lived in that bleak land.

I have been fascinated for a long time by the confluence of cultures. As a young newspaper reporter in Africa almost twenty years earlier I had watched Kamba warriors ride into battle on bicycles with bows and arrows slung across their shoulders; and I had been no less intrigued by African women, dressed only in a brief skirt and a few beads, suckling babies in the stifling heat of a grimly formal colonial courtroom while bewigged and black-robed judges and barristers disputed impenetrable legal niceties. On this trip to the Arctic I had seen contrasts even more startling. For example, we had lunched on a drill ship on T-bone steak and strawberries flown on to the ice by huge Hercules cargo planes, and viewed the ship's pantry filled with steak, fresh fruit, freshly baked bread, and a huge elaborately glazed and decorated fish, the handiwork of a proud chef. Yet, not many miles distant, the Inuit and Métis hunters of Tuktoyaktuk still dug a tunnel thirty feet below ground into the permafrost to store whale blubber, snow geese, fish, and caribou meat. The drill ships' positions were fixed to an accuracy of three hundred feet in the vastness of the Arctic by a signal from a satellite to a computer on the bridge, involving technology so sophisticated that even a ship's navigating officer could not explain it. In the nearby hamlet, native hunters pursued beluga whales with harpoons in an open boat and chased polar bears for their meat supply.

On board an icebreaker, drinking coffee with seamen in their mess, I watched a live CBC documentary exposé about Mafia-type crime in Toronto. Every few minutes the color TV screen disintegrated into a blur of horizontal black and white lines as the ship smashed into the ice, only to return to focus moments later for further revelations about crime godfathers. What, I wondered, did the Inuit—light years away from the Toronto underworld—make of it all? On a later visit I saw Inuit children eating reindeerburgers and ice cream at a greasy spoon in Tuktoyaktuk while they watched the hard-sell antics of a Vancouver automobile salesman dressed as Superman. A little girl of about eight years, her world circumscribed by hostile nature, heard that my home was in Vancouver and asked if that was near Inuvik.

Until my trip to the Arctic in 1977, I had scarcely been aware of Dome Petroleum, but since then I have watched it with growing fascination, noting its corporate maneuvers, each more complex than the last. Over the years, Dome executives gained a reputation as adroit promoters, the smooth-talking snake-oil merchants of the Canadian oil patch, able, with equal facility, to extract enormous sums of money from Ottawa and from the investing public.

For a brief time before 1982, Dome was a corporate Camelot. Whether it was influencing politicians, sweet-talking financiers, or exaggerating its accomplishments, it had an energy, a vision, and a single-mindedness that was unmatched among companies in Canada. Skeptics kept saying the search for oil in the Beaufort was a pipe dream, all promotion; an industrial empire that existed largely on paper and in the minds of its proponents. But it wasn't all talk; by 1981 Dome *had* a gigantic operation at the rim of the world, with the largest corporately owned marine fleet and the biggest private air force in Canada.

To date, not one drop of oil has been sold from the Beaufort and commercial production this decade is questionable. But a lasting impression has remained with me after several visits to the Arctic, a feeling of inevitability that oil *will* be extracted from the North, although clearly there are many complex issues still to be addressed: the possibility of an environmental disaster from a major oil spill, technical challenges, transportation choices, and regulatory details yet to be negotiated.

I realized when proposing this book to my publisher early in 1981 that it would be like taking a snapshot of a hurdler in full flight, so swiftly changing were Dome's activities. I had no idea that Dome would suddenly switch to corporate pole-vaulting with its takeover of the Hudson's Bay Oil and Gas Company, and fall ignominiously at the moment of its greatest glory. The disastrous takeover and its many ramifications— one of Canada's great business stories—occupies a large part of this book. I had not intended to devote much space to the company's financial maneuvering, but as the weight of its debt imperilled its survival, Dome's relations with its shareholders, its bankers, and—perhaps most importantly—the federal government, became the crucial part of the company's story.

This is certainly not a company-sponsored book, and the manuscript has not been seen by anyone at Dome. In October 1982, when I was nosing around Calgary, Jack Gallagher issued specific instructions to his senior staff that they were not to discuss the company's financial crisis with me. But such obstacles merely make an author's job more challenging! Even before the news turned sour for Dome, Gallagher was less than enthusiastic about the prospect of a book being written about his company. He felt then, and now feels even more acutely, that Dome gets far too much publicity. Still, he agreed, with reluctance, to cooperate in granting me interviews and in making it possible for me to visit the Arctic operations for an extended period. Although the demands on his time were so pressing that gaining access to him was never easy, he was always, in true

Gallagher style, exceptionally courteous. Bill Richards is a much more willing—and entertaining—talker, but he too was busy fighting fires. In the senior echelons of the company, Gordon Harrison, a thoughtful and erudite man, was by far the most generous with his time. Further down the corporate ladder, I found a willingness by employees to discuss the company and its affairs, and often a surprising frankness. I gained useful insights from people who were proud of Dome but were not reticent about describing its eccentricities and shortcomings. I am also grateful to many people outside the company who shared perceptions of Dome's activities; some are identified in the text, while others preferred to help me anonymously. The only material assistance I received from Dome was logistical: transportation in the North and accommodation at its base in Tuktoyaktuk and on board ships of its Arctic fleet—with no strings attached. For this hospitality I thank them warmly.

I am deeply indebted to Neville Nankivell, editor-in-chief of *The Financial Post*, and Dalton Robertson, the paper's executive editor, for their enthusiasm for this project and their support and encouragement. Without the leave of absence I was given by the *Post* for research and writing, this book could not have been produced. I must also pay special tribute to my colleague and friend John Schreiner, Western editor of the *Post*, who unselfishly encouraged my leave although it added to his own work load. He also offered much sensible advice and criticism. Jane Davidson, the *Post*'s former editorial administrator in Vancouver, was indefatigable and ever cheerful in helping me with many tedious research chores. I am responsible for any errors of fact, and all editorial opinions are mine alone.

DOME

1

ON TOP OF THE WORLD

Chapter One

OPERATION SWAMPY

It was to be the largest takeover the world had ever seen—and Calgary was the appropriate setting. In the raw early spring of 1981 Calgary was noisily consolidating its wealth. In five years the oil city's population had soared over twenty-five per cent, by far the highest growth in Canada, and three thousand people continued to pour in each month to enjoy incomes far above those of the rest of the country. The sound of jackhammers and the wreckers' ball drowned out traffic noise, and there was so much new construction it was almost impossible to find an unobstructed sidewalk. It was a city where steely-eyed dedication to the accumulation of wealth was housed in hostile towers of mirrored glass; a white-collar professional community infatuated with business, in which everybody was either making deals or talking about them with awe.

Nowhere in uppity, ambitious Calgary was the pursuit of wealth embraced with greater enthusiasm than in the Dome Tower on Seventh Avenue. The leaders of Canada's fastest-growing corporation had many reasons to feel content. Dome, above all others in this vibrant city, exuded confidence and success; it scoured the world for its talented staff and worked them relentlessly, and they loved it and prospered handsomely. They were bright and they knew it, and they sometimes let other people know it too.

One of Dome's most dazzling stars occupied a spacious corner office on the thirty-third floor, decorated incongruously with Japanese dolls and masks, a pair of handsome Western bronze horses, and a wood carving of a red-nosed drunk. Bill Richards, a former small-town Manitoba lawyer who had become Dome's president through unceasing work and a mind of machine-gun rapidity, enjoyed a splendid view of the distant Rockies. There was a predatory glint in his eye as he called in three top lieutenants with a brusque command, "Bring in your hit list." Richards was going hunting.

As he glanced over a list of ten potential takeover candidates, one prey stood out clearly as the most desirable: the Hudson's Bay Oil and Gas Company, known familiarly in Canadian oil industry circles as HBOG (pronounced H-Bog). Richards' appetite had already been whetted. Some weeks earlier, killing time between meetings in the Bonaventure Hotel in Montreal, he had settled down to read the HBOG annual report and other intelligence gathered by his staff. This confirmed his view that HBOG would indeed be a prize: it had massive oil and gas reserves with lands nestling alongside Dome's own properties, and a rich annual cash flow, to say nothing of a noble lineage which it could trace back to lands granted to the Company of Gentleman Adventurers Trading into Hudson Bay in 1670. Perhaps the most significant attraction was the fact that HBOG was paying as much as sixty per cent of its gross earnings in taxes. If Dome could take control of HBOG, that rich cash flow could be diverted directly into the company's corporate coffers and kept out of Ottawa's clutches. Because Dome had been spending huge sums each year in its aggressive search for oil and gas, it had accumulated $1.5 billion in tax credits, far more than it could use. As a result, Dome not only was able to shelter its own income from taxation, but could also shield the income of any profitable operation it acquired—such as HBOG.

A secret scheme was soon under way to take over the Hudson's Bay Oil and Gas Company, code-named "Operation Swampy"—an awful pun on the word "bog". Nothing better illustrates the audacious ambition, the ruthlessness, the go-for-broke gambler's instincts, and the political finesse of Dome than this takeover. At one stroke it made Dome, a mere thirty years old, the biggest oil company in Canada. At the same time, the incautious HBOG entanglement plunged Dome so deeply into debt that its very survival was imperilled.

It is easy in retrospect to deprecate overly eager business adventures that turn sour. It is never simple though, after the event, to recapture the seductive allure of an economic boom, and the accompanying political climate that influenced those making investment decisions. Dome's hothouse growth through the boom years of the late 1970s had been achieved mainly by takeovers, many executed with a brilliance few Canadian companies could approach. People lucky—or perceptive—enough to have invested in Dome early had done exceptionally well: a block of one hundred shares of Dome stock bought for $380 in 1954 was worth $120,000 in May 1981. The company was on a lucky run, and its executives weren't about to pull back.

Dome's HBOG entanglement and consequent brush with disaster was set in train in Ottawa several months before Richards called in his lieutenants, on the evening of October 28, 1980, when the National Energy Program was tabled as a companion to Finance Minister Allan MacEachen's budget. Much later, one of the bureaucrats behind the policy conceded privately that Ottawa had set the climate in which Canadian companies could become greedy by taking over foreign-controlled firms for whom investment in the Canadian oil industry was no longer so attractive. The NEP reflected the leftward-leaning, nationalistic thrust of the Trudeau government which had regained power from the Conservatives of Joe Clark on February 18, 1980. Cooling their heels during their nine months out of office, Trudeau and his caucus had ruminated on the road that Canada should take in the 1980s. During that period of enforced reflection, the thrust in Liberal Party thinking was greatly influenced by two men who argued successfully for interventionist economic policies: Marc Lalonde, who was to be given the energy (and later finance) portfolio, and Herb Gray, who became Minister of Industry, Trade and Commerce.

The new policy had as its objectives the increased Canadianization of the oil industry, and the achievement of energy self-sufficiency through conservation, more determined development of Canada's frontiers,

and the building of new tar sands plants. It was also a
determined effort by the federal government to divert
the large revenues flowing from oil production away
from the province of Alberta and into its own treasury.
In the eyes of the oil industry, one of the National Energy Program's more offensive provisions was the right
of the Crown corporation PetroCanada to seize twenty-
five per cent of oil and gas fields on federal lands—
areas off the east and west coasts of Canada and in the
Arctic. It also imposed a level of taxation that the industry found unbearable; the subsequent decline in exploration activity in Alberta, Saskatchewan, and British
Columbia was dramatic. Oil-rig statistics tell the story:
in Alberta alone there were 390 drilling rigs operating
in January 1981, but by April 1982 this number had
dropped to only 60. Rigs no longer able to find work in
Canada headed south for greener pastures in the United
States; hundreds and then thousands of professionals
fled from Calgary, the heart of the oil industry; and
service industries went into serious decline, sales falling from $1.4 billion in 1980 to $851 million in 1982.
Almost every aspect of industrial and commerical activity was affected: camp caterers, equipment suppliers,
myriad oil-field servicing companies and a charter airlines all saw their recently booming business seriously
eroded.

The NEP was to become an emotional focus of continuing Alberta disaffection with Central Canada and
what were seen as the policies of insensitive Liberal
politicians, not one of whom had been elected west of
Winnipeg in the 1980 general election. The obvious and
widespread rift between East and West, exacerbated by
the NEP, saddened and alarmed many visitors to Alberta. Nothing they had read about Western alienation
prepared them for the ferocious rhetoric they would
hear whenever the policy was mentioned. Bumper
stickers and cap badges flourished, much of the frustration being directed at Ottawa's progeny, Petro-
Canada. Typical of this mood was a bumper sticker on
a huge Buick that stated: "I'd rather push this thing a
mile than buy gas from PetroCan."

Richards was one oilman who shot from the hip; he
labelled the NEP, in a memorable phrase, "confiscation

without compensation." The stock market took note. Dome's share value, along with that of other companies, dipped disastrously, and Richards was in hot water not only with the government but with his peers in the oil industry. Ruefully he admitted later, "The government didn't receive it too well. I think one learns from these things that what you say may be correct but it doesn't always pay to say it." By contrast, the diplomatic Jack Gallagher, Dome's geologist chairman, who had headed the company's remarkable growth from a one-man wildcatter to an industrial giant in less than half a lifetime, flew to Ottawa for a quiet chat with Lalonde. The company built patiently by Gallagher was suddenly in serious trouble.

The NEP linked government exploration incentives with Canadian company ownership: the most generous grants, 80 per cent of exploration costs, would be awarded to companies owned more than 75 per cent by Canadians; those owned less than 50 per cent by Canadians would get only twenty-five cents for each dollar spent. Dome, considered the most nationalistic of Canadian oil companies, whose Beaufort sea drill ships had become the symbol of daring frontier oil exploration, was in fact only 35 per cent Canadian-owned. Unless something was done to change the legislation—or change Dome—many millions in potential grants could slip through the company's fingers. A previous system of tax concessions to stimulate frontier exploration, known as "super-depletion"—or more familiarly "the Gallagher amendment", after the man who had quietly and persistently pressured the government to have it passed—had been scrapped by Ottawa about a year before, and the company, like others in the oil industry, had been on tenterhooks awaiting details of the new grants system. Dome had received no hint of the new key provision regarding Canadian ownership. Now there was consternation in the Dome Tower; the company's fleet of four red-and-white drill ships, wintering at McKinley Bay on the Tuktoyaktuk Peninsula, was due to break out through the ice in seven months to start the important 1981 Arctic drilling season. Without incentives from Ottawa, Dome could not possibly afford to foot the $240 million bill. Already by the fall of 1980

it had moved $50 million in supplies to the Beaufort
and so was "hooked" for the next year.

The company, ever pragmatic, was quick to spot an
ingenious solution to its problems. By the morning following the NEP announcement, it had hatched the outlines of a plan: in future the company's drilling would
have to be done through a "Canadian" subsidiary. Insiders say that Richards, who has an unusually nimble
mind, conceived the Dome Canada idea as he paced up
and down in front of the TV set on budget night. Typically, he refuses to claim sole credit for the idea, saying
merely that it emerged as the result of discussions the
next day. Almost immediately Dome began a series of
fascinating and rapid corporate maneuvers, code-named
"Operation Amoeba", that saw the successful launching
just two months later, on January 30, 1981, of Dome
Canada, a new exploration company with assets of $842
million.

Dome Canada was wildly successful, and Dome's star
never sparkled more brightly than when about sixty
thousand Canadian investors, tantalized by the dream
of potential riches drawn from beneath the Arctic waters,
subscribed $460 million, the largest amount ever raised
for a speculative venture by a Canadian company. It
was a prodigious financial undertaking that involved
practically the entire Canadian investment industry,
and earned the brokerage firms involved about $25 mil-
lion in commissions for just a few months' hectic work.
Dome tapped the Toronto investment firm of Pitfield
Mackay Ross Ltd. to be the principal manager of the
big share offering. It would be Pittfield's job to advise
on the structure of the new company, provide Dome
with market advice, and ensure that the new investment vehicle would be fashioned to find favor with
investors and also to pass scrutiny from various regulatory bodies. Pitfield warned the Dome board that various financial institutions were worried that Dome
Canada shares would not be a legal investment for life
insurance companies, pension funds, and similar institutions, because they are not permitted to make major
investments in companies that have no record of earnings. This was no problem: Dome found an existing
subsidiary with an earnings record, blew the dust off

its charter, and gave it a new name, and that difficulty was speedily overcome. There was an even more delicate matter to be arranged. Dome Canada's potential allure with the investing public depended on the federal government's generous exploration grants. Although *promised* in the National Energy Program, legislation to give effect to these grants had not even been tabled in the House of Commons when Dome and its financial advisers were fashioning Dome Canada. Dome needed to hustle because of the danger of having to delay its 1981 Arctic operations and could not afford to wait out the interminable gestation period of Parliament (in fact the legislation did not become law for almost fifteen months). The issue was settled in a manner that signalled to anyone who ever doubted it the symbiotic nature of Dome's relationship with Ottawa's corridors of power. The Dome Canada prospectus, issued in March 1981, contained a remarkably reassuring letter to Gallagher from Marc Lalonde saying that once he was given parliamentary discretion he planned to use it in Dome Canada's favor in respect of the promised grants.

Dome Canada was hailed by Lalonde, with political hyperbole not appreciated in Calgary, as evidence that the Canadian oil industry had accepted and could work comfortably with his new energy program. Lalonde was on the phone almost daily to Dome as the launching date approached; and he flew to Alberta especially to appear with Jack Gallagher when Dome Canada was unveiled at a press conference. In private conversations Lalonde was voluble in his enthusiastic praise of Dome, much to the irritation of executives at PetroCanada, who felt their own less glamorous activities were being disparaged by an energy minister who had fallen victim to the famed persuasiveness of Jack Gallagher.

Dome's tacit approval of the hated energy program was scorned by other less flexible members of the Calgary oil patch, who felt that Dome had sold out to the enemy. But Richards was defiant: "There was a lot of talk in the industry at the time—maybe we ought to try to defeat the NEP. Well, we couldn't afford the luxury of sitting around in the fond hope that we might be able to defeat the NEP, because even if we succeeded it might have taken another year before the government got around to a new [incentive] regime. In the meantime we had $400 million or $500 million worth of equipment involved in an on-going program in the Beaufort Sea. So the policy we took is, 'To hell with it, whether we like the NEP or we dislike the NEP, it is there, so let's try to make it work.' You cannot spend all of your life engaged in guerrilla warfare.

"There was no question the NEP was popular with the man in the street, and the notion of thinking that you could reverse a policy which is acceptable to the vast majority of Canadians is just ludicrous in my view. It may be all right to tilt at windmills if you have nothing to lose, but when you have one of the biggest investments in your company at stake and what is perhaps one of the biggest exploration programs undertaken anywhere at stake, you can hardly enjoy the luxury of climbing on your soapbox and ranting and raving."

Since the new company would be spending considerable sums on exploration in its early years, it needed some continuing income. Dome decided to transfer to the new offspring half of its own 47 per cent interest in TransCanada PipeLines, then worth $250 million, and to put in $158 million in cash. It also arranged for Dome Canada, which it would manage, to receive a $225 million loan that it had previously negotiated with the Japanese National Oil Company for exploration in the Beaufort Sea. Dome got 48 per cent control of the new firm, and the Canadian investing public, which subscribed $460 million, obtained the remaining 52 per cent. An essential part of the arrangement between Dome Canada and its corporate parent was the undertaking of Dome Canada to carry out substantially all of Dome Petroleum's exploration work in Canada for a minimum of three years, in return for which it could earn up to a half-interest in most of the parent's Canadian exploration lands.

The Dome Canada share offering was the second most successful in Canadian history, coming very close to the $487.5 million raised in June 1979 by the British Columbia Resources Investment Corporation. A total of thirty-six investment firms were involved in selling the stock. In late February 1981, Dome's corporate jets scurried across the country with company officials and senior members of the underwriting teams who made enthusiastic sales pitches, known in the brokerage industry as "dog and pony shows". More than two thousand stock salesmen, investment managers, and big investors got personal briefings on the new offering.

There was an unusually keen demand for Dome Canada shares in the Mackenzie Delta area of the Northwest Territories. In Inuvik, a service center of about three thousand people, just over one hundred miles south of the Beaufort Sea, many local businessmen have much riding on the success of Dome's Arctic operations. Since there is no stockbroker in Inuvik—in fact there isn't one within a thousand miles—the local chamber of commerce asked Dome to make special arrangements to permit local investors to get into the action. The company's response was rapid and much in character: within a week Dome Canada shares were on sale at the local branch of the Canadian Imperial Bank of Commerce, without commission. The first fifty thousand shares were snapped up within twenty-four hours, and more shares were made available the following week to meet demand not only from Inuvik but from other remote northern communities as well. There were a total of 480 applications for $780,000 worth of shares, about ninety per cent of the money used for the purchases coming from savings accounts. This figure does not include shares of Dome Canada bought earlier by local businessmen, who invest regularly through brokers in southern Canada. About 2,800 of Dome's employees across Canada were similarly enthusiastic about the new Canadian subsidiary and paid $30 million to buy about six per cent of the new public float, giving them a strong vested interest in the success of the new company. Not everyone got carried away, however: one Dome financial officer shunned the new venture like the plague and a retired top Ottawa mandarin with a superb knowledge of the oil and gas industry resolutely resisted risking his own money.

Those were euphoric days in the Dome executive offices. The company seemed to have acquired the Midas touch: it was the darling of Canadian investors; it had cemented its friendship with Ottawa by speedily adapting, as no other company had done, to the spirit of the National Energy Program, thereby surely earning much future goodwill in Ottawa's corridors of power; and it was besieged by the elite of Canada's pinstriped banking community fighting eagerly with each other, it sometimes seemed, to lend Dome the most money.

At that time the strictures of the NEP had not yet seriously slowed the frantic activity of Calgary, and the oil capital was still booming. The Bank of Montreal was building a $100 million, 43-story office complex, the city's skyline was littered with cranes, and an incredible $2.4 billion worth of construction permits were issued that year. At the recently expanded airport (where a PetroCanada oil pump nods quaintly near the main runway), no fewer than 126 private executive jets were parked, the greatest number in any Canadian city. Oilmen were constantly on the move between Calgary and the oil meccas of Dallas and Houston; where once they had made pilgrimages to the financial centers of Bay Street and Wall Street, they were now courted on their own turf. Everyone, it seemed, was talking multi-million-dollar business deals, and fortunes were made quickly by the brave. One story is illustrative: a brokerage firm hired a young man in his mid-twenties scarcely out of university, and within three months he was working on nine deals—mergers, takeovers, and financings—involving millions. It was in this intoxicating atmosphere that Bill Richards, heady with the triumph of forming the new Dome Canada, called for the secret "hit list" and began plotting the brilliantly audacious assault to capture the Hudson's Bay Oil and Gas Company, whose handsome oil and gas production, spectacular land holdings, and rich cash flow presented a tantalizing target for what would be the biggest takeover the world had then seen.

Chapter Two

A PYRRHIC VICTORY

Operation Swampy—the bid to capture the Hudson's Bay Oil and Gas Company—was mounted like a military campaign. There were two distinct operations: the first, a brilliantly executed commando assault; the second, a disastrous encounter resulting in months of trench warfare that sapped Dome's resources and saw the humbling of a seemingly invincible warrior.

A quick reconnaissance of the terrain determined the first battlefield. HBOG was 52.9 per cent owned by the giant U.S. oil and coal company Conoco Inc., with Canada's Hudson's Bay Company controlling 10.2 per cent. To get its hands on Conoco's controlling interest in HBOG, Dome's planners hit on a brilliant and unorthodox outflanking movement. Its maneuver was simple: it would offer to buy, at a premium price, between thirteen and twenty per cent of Conoco shares on the open market, thus threatening Conoco's autonomy, and then trade these back to a desperate Conoco in exchange for its holdings in HBOG.

At first sight, the prospect of Dome trying to force its unwanted attention on Conoco seemed farcical: Conoco, the tenth-largest U.S. oil company, towered over the Calgary upstart. At the end of 1980 Conoco had assets in excess of $11 billion and had made a profit of more than $1 billion, while Dome, by comparison, had assets of $5 billion and had made $287 million profit. Conoco was surprisingly vulnerable, however. Its shares were so widely held that no individual or company owned more than three per cent. Even more to the point, its shareholders, as it turned out, were unimpressed by

Conoco's recent performance and were embarrassingly eager to sell their interest in the company.

The initial reconnaissance was made on March 12, 1981, by Jack Gallagher, who happened to be in Stamford, Connecticut, for a board meeting of Texasgulf, of which he is a director. Conoco's headquarters is also in Stamford and he called Conoco chairman Ralph Bailey to ask if he could drop by for a chat. At first it seemed little more than a courtesy call, but as he was leaving, Gallagher said he would like to call again to discuss HBOG. In late April he was back, this time accompanied by Richards.

Much was made later of this meeting which ended in serious misunderstanding. Gallagher and Richards did all the talking, since Bailey said he was there merely to listen, presumably on the advice of lawyers. The visitors from Calgary suggested that Conoco's interest in HBOG was now clearly worth less to the U.S. company than it would be to Dome, since Canada's NEP had severely hampered foreign-controlled oil companies operating in Canada. Richards made the point that sometimes the best purchase a company can make is buying back its own stock from the public. If Conoco re-purchased its own shares and then retired them from the market it would immediately enhance the value of its fewer remaining shares. The Dome visitors said they had hit upon an idea that would benefit both companies, characteristically involving a little sleight of hand with taxes. If Dome bought HBOG shares directly from Conoco, the U.S. company would have to pay a capital gains tax. On the other hand, if Dome made a tender offer for Conoco stock and then exchanged that stock—which Conoco could then retire—for Conoco's interest in HBOG, Conoco could escape taxation. Dome's lawyers believe that there was a precedent for this in U.S. law. Bailey was not impressed by the important tax argument. Later, when the confrontation with Dome became heated, he suggested that Gallagher and Richards idemnify Conoco if Dome's tax interpretation was faulty. They didn't take him up on the offer.

But at that crucial meeting Bailey simply didn't respond to the suggestion made by his visitors. The Calgary oilmen misinterpreted his silence; they were sure that Bailey was giving them an implied agreement to proceed with their offer, while being punctiliously correct and discreet. Gallagher flashed the smile that had illuminated boardrooms and charmed cabinet ministers and thought he had an unspoken deal. Bailey, who subsequently refused to discuss the matter, apparently felt that he had given the Dome executives the cold shoulder and that nothing of significance had occurred at the meeting.

Consequently, he was outraged on May 5 when Dome made a public offer of U.S.$65 a share for between thirteen and twenty per cent of Conoco, then trading in New York at about U.S.$50. Robert F. Greenhill, managing director of the investment banking firm of Morgan Stanley and Company which was advising Conoco, called Gallagher and told him he was crazy, and a spirited but unsuccessful attempt was made to block Dome's scheme through the U.S. courts. Gallagher, who dislikes hostile takeovers, said Dome went ahead because it appeared that it would be a friendly situation. "And then the investment dealers and lawyers got hold of Bailey and they could see a fee in it for themselves, win or lose, and they convinced him to fight. So that delayed the takeover and created all the animosity that developed during that period; I think if we had recognized that problem we probably never would have moved on it," he said.

For all his amiability, however, Gallagher is made of tough material. Bailey and Gallagher exchanged terse letters in which Gallagher hinted strongly that if Bailey had any ideas of selling Conoco's HBOG interest to anyone else (Calgary's Nova Corporation, a Dome rival, was also sniffing around), he had better forget them; if he did that, Dome might seek to swallow up Conoco itself. It's an academic question now whether Dome would ever have been able to muster the financial muscle to follow through with this threat. Richards maintains it could have, not on its own, but certainly in association with others. He says that throughout the entire exercise Dome had a preferred route and a fallback position; if it had been obliged to take over Conoco it would have been able to do so.

Bailey wrote to Conoco shareholders that Dome's

move was "nothing less than an attempt to make the
company sell its interest in HBOG at a price which is
grossly inadequate and on the basis of a series of transactions involving alleged tax benefits which tax counsel
advise are highly questionable." He added, "We believe
it important that our government consider what action
should be taken so that U.S. citizens are not victimized
by Canada depressing the value of their assets and then
discriminating in favor of its nationals acquiring those
assets at depressed values." This was a brave bluster
by an executive who was shooting blanks. In the event,
Dome's offer was dramatically oversubscribed: the Canadians were offered fifty-one per cent of Conoco, although they bought only twenty per cent, and Bailey
realized that he would have to come to terms, since at
that point Dome held what Richards called "a balance
of fear". "We had control of a block of Conoco stock and
that wouldn't have been a very desirable position from
their standpoint," he said later.

Armed with twenty-two million Conoco shares, for
which Dome had paid U.S.$1.43 billion, Richards flew
from Calgary to Stamford on Sunday, May 31, 1981.
As the Dome president and his aides drove up to the
Conoco headquarters, another Canadian, Charles
Bronfman, vice-chairman of Seagram, was already there
with Bailey, watching them unseen through copper-
tinted one-way glass. Bronfman's presence was coin-
cidental and had nothing to do with Dome's activities;
he was intent on making his own deal for a major stake
in Conoco. However, Richards and his party were se-
questered in a utilitarian meeting room as they worked
on the final details of their deal. The atmosphere was
oppressively hostile and a security guard even accompanied them to the washroom. Apparently these precautions were taken so that the Dome party should not
know that Bronfman was in the building. Probably
Conoco was afraid that Richards might decide to sell
his newly acquired Conoco shares to Bronfman at a
quick profit if he became aware of his interest in the
U.S. company. Discussing this much later, Richards
laughed uproariously; he had no clue, he said, that
Bronfman was around. The Bronfmans had, in fact,
considered approaching Dome earlier after reading of

the success of its offer for Conoco stock, but had decided instead to approach Conoco directly to try to acquire its own interest in the U.S. company.

At the time, the Dome contingent had a different worry: they were afraid that their private discussions might be bugged by their hosts. Spying on the competition for drilling results is a time-honored oil-industry practice when sums worth only a small fraction of those being discussed that day in Stamford are involved. They had no particular reason to suspect that Conoco would engage in electronic surveillance, but with stakes in the billions they were naturally not prepared to take chances, and so, in true spy-movie fashion, they were careful to walk around in Conoco's beautifully landscaped gardens while they discussed hard numbers. Earlier, in Calgary, Richards' assistant, Steve Savidant, had sought advice about possible bugging from Dome's own security chief. "Our security people said, 'We don't have any idea whether they would do it [use bugs] or not, but we should tell you if they wanted to, they have the capability and they have the sophistication. If you are in a room and you really want to have a serious conversation and you are concerned, turn a radio on.'" So when they needed to talk to Gallagher, who was at his weekend cottage on Lake Windermere in southeastern British Columbia, the Calgary oilmen drove into Stamford and booked into a hotel. "We booked into a room—any old room—and got on the phone to Jack. I guess we were in the hotel room for maybe forty-five minutes. But we left the Conoco building really out of an excess of caution," said Savidant. The Dome party also took along an extra pilot to guard their executive jet while it was in the Conoco hangar. They were concerned that Conoco might bug their aircraft to monitor conversations as it flew back to Calgary if negotiations were not concluded that day. "We were not paranoid, but we were dealing for $2 billion. We were on their turf, we were in one of their limousines, their building, their conference room, and using their phones and their hangar out at the airport. When the stakes are that high, you can't take chances," said Savidant.

Conoco demanded a high price in return for the surrender of its share in HBOG: the American company

sought the twenty-two million shares for which Dome
had paid U.S.$1.43 billion, plus U.S.$245 million in
cash, for a total price of U.S.$1.68 billion, fifty per cent
more than the amount at which Conoco had been valued
by the stock market just weeks before. Intense negotiations continued all day Sunday and throughout the
night, concluding in agreement at seven o'clock on
the Monday morning.

It was a group of weary but elated Dome executives
that slipped into armchairs on board the company's lux-
urious $8 million Gulfstream II jet (a conveyance cost-
ing four thousand dollars an hour to operate and known
in the aviation trade as a "royal barge") for the journey
back to Calgary. During the flight the executives enjoyed a drink or two, and from 38,000 feet Bill Richards
called up HBOG President Dick Haskayne in Calgary to
invite him to lunch that day. (Gallagher made a similar
call to chairman Gerry Maier, and the four men met
for an amicable meal.) Richards was in high spirits
when he got back to Calgary and told reporters cockily,
"I want to make one thing perfectly clear: we have no
intention of bidding for Imperial Oil."

For obscure legal reasons the final documentation
had to be handed to a Conoco representative on Canadian soil, and so a young female paralegal employee
of Conoco flew back to Canada with the Dome executives. "We were all kind of silly, joking a lot on the way
back. She must have thought we were a strange lot,"
said Savidant. The formal documentation was handed
over in Dome's hangar at Calgary airport, and the Conoco representative returned to Stamford on the next
commerical flight.

When the Dome assault on Conoco was announced
on May 5, it came as a total surprise to the executives
on Second Street S.W. in Calgary who were busily engaged in running the Hudson's Bay Oil and Gas Company. (They learned with sadness later that Conoco,
their American majority owner, had already made another, earlier, attempt to sell their company, that time
to PetroCanada; but Wilbert Hopper, chairman of the
Crown corporation, had declined because the asking
price was too rich.) It was difficult for the HBOG officers
to comprehend a takeover of such magnitude, involving

almost double the sum previously paid in a takeover battle anywhere. But they were hardly paralyzed by the news—within ten days of Dome's opening salvo, the HBOG executive officers had managed to secure "golden parachutes" for themselves. HBOG entered into an agreement with all of its officers which ensured that whatever happened in the coming battle they would be well provided for. If they quit or were fired within two years of the change of control of the company, they would get a lump-sum cash payment equal to three years' salary plus the amount of any bonus they received in the fiscal year prior to termination. Those officers who remained with HBOG for two years after the change in control would get an immediate bonus of fifty per cent of the above formula.

Employees working at a lower level for HBOG did not enjoy such splendid arrangements, but all was well at the top. In 1980, for example, HBOG chairman Gerry Maier earned $194,461 in salary, fees, and bonuses. Dome wanted Maier to remain after the ultimate merger of the two companies, but he departed in some bitterness and a few months later became president of Calgary-based Bow Valley Industries. He took with him well over a quarter of a million dollars. In all, six HBOG executives resigned rather than work as subordinates to Dome, although Dome tried hard to persuade them to stay. But many HBOG employees remained with the company when it merged with Dome months later, and Dome did make a genuine attempt to see that their career progression was uninterrupted.

For all the brilliance of the assault on Conoco, the acquisition of little more than a half-interest in HBOG placed Dome in a desperately vulnerable position. Even before its attack on Conoco, Dome had a staggering $2.65 billion in long-term debt, and there were whispers from more cautious members of the oil industry that the company would have to confront a day of reckoning. The purchase price for Conoco's HBOG interest was financed entirely by four Canadian banks, so that the Calgary company did not put up so much as a nickel of its own money. As interest rates reached unprecedented heights, repaying this debt drained the company's lifeblood until it had scarcely any money left over to run its day-to-day operations. There were many factors involved in the catastrophe that was to befall Dome, but the brutal increase in interest rates was the single most injurious. In the summer of 1981 rates reached twenty-three per cent and more—levels that not even the most conservative financial manager could have thought possible, since usury of this magnitude had formerly been the exclusive territory of the Mafia.

It became clear to Dome's planners that the company would have to take over HBOG entirely or it would certainly perish. There were two reasons for this. Under the accounting rules which regulate all corporate activities, Dome, although now the major shareholder, was entitled to receive only annual dividends from HBOG, the same as any other shareholder. In Dome's case these would amount to about U.S.$12.65 million, which was nice, but not enough to pay even two weeks' interest on its additional debt. To lay its hands on HBOG's handsome cash flow—a golden egg twenty times bigger— it required 100 per cent control. Also, it needed to sell off parts of the HBOG empire to lighten its debt burden— but it could not do this until it was the sole legal owner.

In postmortem discussions much later, Dome executives agreed they did not move speedily enough, because they did not fully appreciate the urgency of their situation in a rapidly deteriorating economic climate. They also procrastinated because they hoped that good news from their Beaufort Sea drilling operations would drive up Dome's stock price and so strengthen their hand. To complete the second half of the takeover, Dome had to come to terms with the Hudson's Bay Company, which held 10.2 per cent of HBOG stock, yet it could give no hint that it was desperately eager to buy. Richards explained the slow-moving negoitations later: "We were going around trying to pretend that we really didn't need to do the second half [of the deal] to try to enhance our bargaining position, because if we had gone to the Hudson's Bay Company and conveyed a sense of desperate need to complete the balance of it, we might have had a more difficult bargaining position than we did have. It's pretty hard to say how we could have speeded up the process without seeming to exhibit a sense of indecent haste."

Dome first tried to acquire the Bay's interest in HBOG by offering a straight share swap, which it considered an attractive offer since Dome's shares were then trading about twenty-five per cent higher than those of the Bay. But when Gallagher met with newspaper magnate Kenneth Thomson, the Bay's controlling shareholder, Thomson turned him down. Dome then proposed an alternative: it would buy all the remaining HBOG shares by issuing a type of security known as a "retractable preferred" share, which conferred important tax advantages on the holder. Using this method Dome would undertake to pay quarterly dividends for a number of years, and would then "retract" or buy back the shares. It was agreed that Dome would negotiate the details of the deal with an independent committee appointed by the HBOG board of directors to represent all the remaining shareholders.

The committee turned out to be exceptionally hard-nosed horse traders. Amond them was Ian Barclay, chairman of British Columbia Forest Products Ltd., a no-nonsense corporate warrior who sheds fascinating light on Dome's method of conducting negotiations. Many months later he was still bewildered by what had happened.

"It seemed to me that in everything Dome did, a decision was required before you got there. You were in a working environment like a pressure cooker, and until you got used to it it was a little wearying on the nerves. Dome's first stage was almost a bulldozer approach. If you are going to get bulldozed over, that's fine. If you can get through that one, you eventually reach a slightly more calm and reasonable manner of discussing things. One of the things that happened was that different bits and pieces of information seemed to be coming to different players all the time. Either our lawyer would hear something from their lawyer, or our financial person would hear something from their financial person, or one of us may have heard something from Gallagher or another one of us may have heard from Richards...and they weren't all necessarily the same thing. A lot of the discussions were rather fruitless because we all thought we were dealing with the same piece of paper when none of us were."

Was Dome operating a strategy of confusion, and, if
so, what did it stand to gain? Like a highly competitive
athlete it may have been attempting to "psych out" its
opponents. "I never really could figure out whether it
was the way they did business or whether at times the
left hand didn't know what the right hand was doing.
I think it was probably their style, [because] if you talk
to others who have dealt with Dome you get the same
feedback. When Jack said, 'Well, Bill did this and I
really didn't know about it,' was it a technique or was
it a fact? You just didn't know. It seemed hard to believe, when you are dealing with numbers like that,
that there wasn't some communication between the two.
It was a very complex transaction, certainly the biggest
in Canadian history."

Central to the difficulties faced by the negotiators
was the value to be placed on various oil and gas properties. "Because Dome were using the preferred-share
route it required an evaluation of Dome, which isn't at
all easy to do," Barclay says, explaining that part of
the problem faced by his committee was getting a specific proposal from Dome. "Finally we had a meeting
with Dome and we told them their figures and proposals
were not acceptable to us and they said, 'Look, you as
a committee, tell us what you are prepared to recommend, and you know the Bay will accept.' So the committee—and don't forget we had our own engineering
consultants, our own financial consultants, and our own
legal counsel—came up with what we thought was a
reasonable offer, and we met with the Bay and trotted
it out and discovered we were within whiskers of being
in the same ballpark." Dome's opponents drove a hard
bargain, demanding—and getting—a price per share
eighteen per cent higher than Dome had paid Conoco
in a much more buoyant market months earlier.

Dome had still another hurdle to face and it had
darkly ominous implications which no one appreciated
at the time. The independent committee, skeptical of
Dome's ability to meet the financial obligations of the
preferred-share issue, insisted on an insurance policy,
and a sizeable one at that: $1.8 billion in backup fi-
nancing to be invested in gilt-edged securities and held
by the Montreal Trust Company, where it could not be

touched by Dome. Dome sought the money first from the Canadian banks but was turned down; undeterred, it arranged the loan through a 26-member consortium headed by Citibank of New York, which had long been after Dome's business. Each bank in the group needed to be satisfied as to the security of the HBOG assets being pledged as collateral, and so a small army marching with briefcases invaded Calgary. The banks each sent in three or four experts to examine the HBOG oil and gas reserves. So many visitors were involved that a meeting room at the Calgary Convention Centre was provided where they could scrutinize the voluminous documentation.

The details of the deal between Dome and HBOG's minority shareholders were finally completed in a marathon negotiating session at the King Edward Hotel in Toronto that concluded on November 3. Ian Barclay describes it as his forty-hour day.

"I think I was on the phone for about eight hours on Sunday [in Vancouver] and I flew to Toronto on the Monday morning. I went to the King Edward Hotel at about five in the afternoon and I next got back to my hotel room at 2 a.m. on the Wednesday morning. That was just at the end. Hell's bells, other people had been there all Sunday and Monday morning. It was crazy. We always seemed to be meeting on Saturdays or Sundays and we always seemed to be going until the early hours of the morning. I don't know whether it was meant to be the survival of the fittest or what the hell it was."

Just when Dome finally believed it was sailing into protected waters, a new menace appeared like an enemy periscope from an unexpected quarter: Ottawa. On November 12, 1981, Allan MacEachen's second budget was tabled in the Commons. One provision seemed like a lethal torpedo aimed directly at Dome's bows: effective at midnight, a key tax loophole, previously available to investors in share exchange deals and vital to Dome's arrangements with HBOG, was bolted shut like a watertight door. Telephone lines between Calgary and Ottawa glowed hot the next day. High officials in both the finance and the energy ministry were lobbied and Richards talked by phone with Marc Lalonde. Richards and a group of senior Dome people flew to Ottawa

and began a siege of government that was to last about
a week.

"We mounted an intense campaign to make sure that
the politicians and senior bureaucrats understood the
issues and the implications of the budget," says Savidant.

Their prey was tracked down relentlessly. Secretary
of State Gerald Regan, for instance, had been out to the
Maritimes, and Richards, discovering when his government plane was due back in Ottawa, was there at the
airport to greet him with a waiting limousine and a
sales pitch. Finally, the Dome team was called into
MacEachen's office.

"As we walked in, apparently he decided he would
have a little fun with us," says Savidant. "He seemed
to be just glowering at us, with dark clouds over his
head. He had a piece of paper in his hand and he said,
'Have you read this?' I said, 'No, sir, we haven't.' He
said, 'Well, read it.' So we sat down and read it, and it
was the statement he was going to make in the House
of Commons." MacEachen then went into the Commons
to announce that the budget provisions would not apply
to Dome's deal and certain others that had been agreed
on but not carried out when the budget was introduced.
No pope ever gave a more welcome dispensation.

Although it was now able to complete a merger with
HBOG and tap additional funds, Dome's need to reduce
its debt was obvious and pressing; in fact, an urgent
plan to slim down this burden was already under way.
Even before it had concluded the deal, Dome had arranged to sell off part of the company it was still to
acquire, although legally it could not finalize these
transactions until it had complete ownership of HBOG
in its pocket. Several of these sales were to members of
its own "family", like Dome Canada, which agreed to
pay $460 million for a 12.5 per cent interest in certain
HBOG properties in Western Canada. A subsidiary of
Dow Chemical of Canada Ltd. agreed to pay $430 mil-
lion for a similar stake in HBOG. And TransCanada
PipeLines Ltd. was also to buy 12.5 per cent. Richards
told financial analysts in Calgary in January 1982 that
the company intended to further trim its debt by $800
million by selling more assets by the end of the year.

But Dome encountered severe difficulties accomplishing this aim, and as the world economy slipped more deeply into recession, the company's financial plight turned to one of desperation.

Only a handful of key executives knew that a financial time bomb had begun ticking in the Dome Tower.

2

THE RISE OF DOME

Chapter Three

JACK GALLAGHER: "A CLASS ACT"

At 8 p.m. on Wednesday, September 29, 1982, in the plush 56th-floor boardroom of the Canadian Imperial Bank of Commerce in Toronto, Jack Gallagher endured one of the blackest moments of his life, a humiliating defeat of awesome dimensions. He signed a document, forced upon him by bankers and bureaucrats, that would allow them to wrench away control of the company he had built with parental devotion for over thirty years. He was being vilified in newspapers and magazines across the country as a blundering business incompetent, and his loss of personal fortune was scarcely without equal in Canadian history—more than $118 million erased in about twelve months. More than likely his back, seriously injured in a truck crash in South America four decades before, was giving him hell as it often does. After the document was signed, people began drifting towards the doors, while others busied themselves with a press release that still had to be written on "The Dome Bail-Out". Suddenly Gallagher stood up and asked everyone to stay a few moments longer; he just wanted to express his thanks for the hard work they had put in to help Dome with its problems. It was vintage Gallagher, a quietly gracious speech, upbeat and optimistic, and, in the circumstances, totally unexpected. An astonished bank official conceded later that Gallagher's performance, under stress, was "a class act".

One colleague who has observed Gallagher closely says, "I can't believe that fellow. His control of emotion

is incredible. He doesn't sweat; he never betrays his
feelings by wringing his hands or tugging his ear or
pulling his nose. I think he has some special vision of
himself, some feeling of destiny."

Everyone who meets "Smiling Jack" Gallagher is
captivated by him and it has been that way for half a
century. His famous smile is truly dazzling, and even
in his late sixties he is a remarkably handsome man,
a trim, graceful figure with silver hair and a slim moustache of the type that used to be favored by Hollywood
stars several decades ago. Much has been made of his
charm and it is a dour person indeed who is not instantly captivated by a personality that radiates warmth
and goodwill towards all right-thinking people. As many
a junior Dome employee has discovered, there is something seductively appealing about having a highly intelligent, vastly experienced national figure solicit your
views and listen to them with obvious interest. Gallagher's charm is not merely practiced technique, a
salesman's tool to be filed away when not required;
those close to him maintain that it is a permanent man-
ifestation of a genuinely kind personality. They do not
deny, however, that "Smiling Jack" is as tough as old
nails. He pursues his own self-interest doggedly while
waving nationalist banners, and can be ruthless in
business when occasion requires. An oil-industry peer
cautions: "Jack Gallagher isn't exactly a *gentle* man,"
which is a far cry from saying that he isn't a gentleman,
since he undoubtedly is. The saying goes within Dome
that he can fire a person and they will thank him for
it. He is a complex character, whom it is perhaps necessary to view from the beginning in order to try to
understand.

A picture of the young Gallagher is offered by a
schoolboy at St. John's Technical High School in Winnipeg's North End, who wrote this thumbnail sketch
for the 1932-33 yearbook: "Jack Gallagher, 5'10", eyes
of blue, ash blonde hair, modest, retiring air, disarming
smile; a lively advertisement for good toothpaste is J.
G. His popularity with the opposite sex is acknowledged
by everyone except himself. Shy, upright. Business
manager of *The Torch* [the school magazine]."

Dr. Sam Rusen of Winnipeg, a classmate at St. John's,

recalls, "Even then he was a charmer and had that famous smile. He could charm the birds off a tree. He became president of his class in almost no time. He was a good student, very clever, I remember. I lost sight of him after high school and never heard of him again until three or four years ago. Then I saw him on TV, and even before hearing the name I knew it was Jack just from that smile."

Gallagher was an active debater and was also in the school orchestra. Another school colleague, retired Winnipeg photographer John Boratsky, says, "He really was a handsome devil, always turned out immaculately. That was in the Depression, but he usually had a suit on and looked very smart. He was modest too. Just about everybody in the class participated in athletics, and Jack—they called him 'Galloping Gallagher'—was no exception. They referred to him as the 'half-mile prodigy' in the yearbook. He was one of the outstanding athletes."

Gallagher enrolled as an engineering student at the University of Manitoba because he enjoyed designing and building things and wanted to work outdoors. But he soon realized that he would have difficulty in getting a job as a construction engineer during the Depression, and so he switched to geology, encouraged by the decision of Prime Minister R. B. Bennett's Conservative government to set aside a few million dollars to explore and develop Canada's mineral resources in its "Vision of the North".

However, it was Gallagher's working vacations which became the most formative periods of his education. In 1936 he took a summer job with the Geological Survey of Canada, and the deep impressions of weeks spent in the northern Canadian wilderness have remained with him all his life. "It was a tremendous experience for a young man to get in there and see the vastness of Canada and how little it had really been explored," he says. With just one companion he went into northern Manitoba in a single-engine Junkers bush plane, their canoes strapped onto the aircraft's pontoons. The pilot, flying by the seat of his pants, with only oblique aerial photographs for reference, would land on a remote lake and arrange to pick up the two young geologists many miles away at another lake, eight or ten weeks later. Tormented by blackflies and mosquitoes in the days before insect repellents, Gallagher and his companion had to wear hoods over their heads, sew up the side openings in their shirt sleeves, and smear their faces with citronella and pine tar, which made them smell so bad that few insects would bite them as they gathered geological data. In order to avoid insects as he recorded his notes, Gallagher often found it necessary to gather moss from the rocks, light it, and stand over the smoke as he worked. They lived mainly on dried beans, dried peas, moldy bacon, and whatever fish they could catch. But as they travelled through the wild, lonely bush, they named their own lakes and rivers, and it was during this youthful experience that Gallagher's lifelong fascination with the potential of Canada's North was born.

After graduation he went back for a second summer in the North, this time as the expedition leader, with a raise from $2.50 a day to $2.85. "You couldn't spend any money, and anyway I always felt that I really should have been *paying* for the experience and for the exposure to a very fascinating geology." The second year he visited Yellowknife, an embryo community in those days. "There had been some claims staked in Yellowknife the year before. We worked from there north through into Gordon Lake and east along the north shore of Great Slave and north from there in through all those small lakes. We were back in pretty virgin country." The experience was valuable because Gallagher learned to decipher the complex geology of the Precambrian Shield, but, perhaps more importantly, it toughened him so that none of the physical hardships he was to encounter later as a field geologist seemed daunting. On portages he had to carry on his back all his supplies for eight or ten weeks as well as a seventeen-foot canoe. Later, by comparison, the deserts of North Africa and the jungles of the Amazon and Peru, where he had native bearers to carry his load, presented no hardship.

In 1937, although he had never seen an oil well, he joined the Shell Oil Company in California. Realizing that all of his fellow workers had master's and doctor's degrees, he once asked why they had hired him, and was told that Shell felt his experience in the Canadian bush was one of the best starts a young man could get in geological fieldwork. He travelled the world as an oil company geologist for more than a decade but was always determined that once he had sufficient practical experience under his belt he would return to help open up the Canadian Arctic.

In January 1939, Gallagher was transferred by Shell to Egypt, and for a time he was on loan to the British Eighth Army in the desert. The British troops had a serious problem obtaining drinking water, since pro-German Egyptians kept cutting a water line from Alexandria. One evening, driving along a coastal road about two hundred yards from the Mediterranean, Gallagher spotted faint traces of greenery on the side of the desert road caught by the slanting light of the setting sun. It was a chance sighting, because the faint blush of green would have been obliterated in the intense glare of full sunlight. He ordered a hole drilled and found a layer of spongy limestone filled with fresh water, and below that the salt-water table from the Mediterranean. The delighted British Army, following Gallagher's directions, dug a series of shallow pits and were rewarded with a vital new supply of fresh water.

Years later in Calgary, Gallagher heard an interesting sequel to this story from the West German military attaché to Ottawa, who had been with Rommel's Afrika Korps as they pushed the Eighth Army to the west during the desert war. The German officer said that Rommel too was plagued by a water shortage. The Afrika Korps had found the pits dug by the British and had assumed they would be poisoned, but when they tasted them gingerly, they were puzzled to discover only salt water. Apparently as the British had retreated they had deepened the fresh-water pits to allow the salt water to seep up from beneath, and the Germans could never understand why the British had dug all these holes to get salt water when the Mediterranean was only two hundred yards away.

With the outbreak of war Shell folded its exploration operations in Egypt, and rather than moving to the Dutch East Indies as he had been asked, Gallagher joined Standard Oil Company of New Jersey and spent

two years in the desert areas of Egypt and Palestine
and the adjoining countries. When he visited Cairo he
was asked to give a few lectures in geology at the American University. He also spent over six months in the
Sinai desert and was impressed by the millions of years
of geological history evident in the beautiful and distinct weathering colors on rocks of various ages.

Like his exposure to the Canadian North earlier, this
desert experience remained vivid in Gallagher's mem-
ory, and many years later he devised and promoted a
unique if rather grandiose plan to solve one of the problems faced by the troubled Middle East. In 1972 he
wrote to Prime Minister Trudeau trying to interest him
in a project that would involve the irrigation of 120,000
acres of the Sinai Peninsula to create a fertile homeland
for Palestinian refugees. He also raised the idea with
the Secretary General of the United Nations, and various other world figures. Like all Gallagher proposals
it did not lack ambition, for it involved the building of
two nuclear-fuelled power and desalinization plants at
a cost of between $1.5 billion and $2 billion, a modern
housing development, a fertilizer complex to use the
potash and other natural resources available in the area,
airports, a highway system, and possibly a trans-
peninsula oil pipeline. The nuclear-power plants could
be supplemented by the billions of cubic feet of waste
gas that were being flared in the oil fields of the Gulf
of Suez and the Persian Gulf. A little naively perhaps,
he suggested that the United States, the Soviet Union,
France, the United Kingdom, West Germany, and Canada should pay the initial cost. The plan, which Gallagher had first suggested in a series of speeches in
1951, came to nothing.

From Egypt, Standard Oil sent him to Latin America, where he searched for oil in the lush jungles of
Ecuador, Guatemala, Honduras, and Panama. Despite
the humidity and the endless rains, he remembers this
as one of the most enjoyable times of his life. It was
during these years that the young geologist honed a
natural talent for leadership which was to remain with
him throughout his career. In Ecuador in 1942, he made
three trips from the valleys of the Andes across the
western mountain range and down to the Pacific Ocean

by foot. The party he was leading cut its own trails, using as bearers natives from the highlands who were not even aware of the existence of the Pacific Ocean. The terrain was difficult and there were no native inhabitants along the routes, so Gallagher had to ration the food. Like a nineteenth-century missionary he recalls, "These illiterate natives had no idea but that we were leading them to their deaths, but in complete faith they stayed with us. And at the end they saw the beautiful blue Pacific Ocean." Gallagher also hired Hibero Indians, better known as head-hunters, as canoemen in the headwaters of the Amazon. "They brought their women with them. We found these people very tranquil; as long as we only had a few workers with us from outside the Hibero area, and restrained those men from going after the Hibero women, we had absolutely no trouble. They were not interested in being paid money but were happy to work for trinkets such as small mirrors, beads, and brightly colored cloth, which they could use to adorn their womenfolk." Travelling was easy in those days for Gallagher, who carried only a toothbrush and pajamas, and bought local clothes and work boots.

Standard wanted to get oil out of Ecuador not only for commerical sale, but also to supply the U.S. Navy if the Panama Canal was put out of action. "Unofficially, I guess, we were really working for Donovan's oss." This was General Bill Donovan's American Office of Strategic Services, the Second World War forerunner of the Central Intelligence Agency.

During this time Gallagher developed another important new talent: the ability to seek out and beguile government leaders. In 1944 he went to Guatemala, which was considered one of the most stable governments in Latin America, and sought permission to explore the northern part of the country on Standard Oil's behalf. He was turned down by a junior minister. Undeterred, he sought out the Guatemalan external affairs minister, who arranged a fifteen-minute meeting with the president, General Jorge Ubico, a dictator who had ruled with severity for fourteen years. By this time the 28-year-old Gallagher could speak a fair amount of Spanish, and the meeting in the dictator's own language lasted for an hour and a half. The tough general warmed to the young Canadian and asked a great number of questions. He wanted to know why geologists studied rocks in the mountains and yet seemed to do their drilling in the valleys, and Gallagher obliged with an introductory lecture in geology. After the meeting the general ordered letters prepared for Gallagher addressed to each of what Gallagher called his "hefty politicals" around the country, telling them to give him mules or canoes or whatever other help was necessary. "He gave us one year in which to look over the country and he did not charge me a penny for it, although he knew he could because he was dealing with Standard Oil of New Jersey and not Jack Gallagher. Shell had been in there a few years before and they had to pay $800,000 just for the right to explore." As his guest departed, the dictator said in perfect English, "Mr. Gallagher, it was very nice of you to come in."

Cheered by Ubico's blessing, Gallagher went off on a field trip, but returned three weeks later to find that the helpful dictator had been thrown out of office and replaced by a *junta* of three. Luckily, he had met one of the heads of the new regime previously, and so retained access to the palace and to certain aerial photographs he needed. A second revolution some months later was much more severe, although Gallagher was never in any personal danger. Although impressed by Guatemala's great oil potential, he advised his bosses at Standard Oil not to undertake exploratory work because he felt the new government was Communist-inspired, which turned out to be correct.

Gallagher's reluctance to discuss another interesting incident in Latin America even after forty years reveals a streak of extreme caution in his nature and his acute political sensitivity. In 1942, while working in the northern part of Ecuador, he came across caches of arms in an area where the Colombian Oil Company, staffed partially by Japanese, had worked earlier in close contact with German landowners. When asked about this now, he will merely say circumspectly, "We did come across some indications of caches that were helpful to the authorities to know were there, but we never did come in contact with the so-called Colombian Oil Company personnel who moved back north into Colombia as we moved north." Pressed for details, he cautions, "We are dealing with the Japanese in our business and I'm very reluctant to talk about that [incident] because it can boomerang on us. The Japanese have been excellent partners [as investors] for years in the Beaufort and very supportive, and I wouldn't want to do or say anything that might upset them. To me Dome is far more important than anything else; it is part of my life." He is unimpressed by the argument that such sensitivites must surely have evaporated after four decades.

Gallagher carries more than just memories of his years in South America. He had a chilling escape from death in a truck crash in the Andes when his driver passed out with altitude sickness at about 16,000 feet. The road was narrow, without a guard-rail, and to the right was a 6,000-foot precipice. Gallagher reached over and wrenched the steering wheel to the left, but the truck crashed into a rock wall and Gallagher suffered a serious spinal injury which caused a loss of feeling in his left arm and leg. It was ten days before he returned to civilization and could be examined by a doctor in a small Peruvian town. The doctor concluded that Gallagher was only suffering from muscle spasms, and so he carried on his field work in considerable pain. It was over a year later that he was X-rayed in New York, where it was discovered that he had fractured three vertebrae which had by then fused on their own. Gallagher says lightly, "I still experience some muscle-spasm problems in the area of the injury, but I'm sure that the problem is not between the shoulders but more between the ears." Those who know him well say that he suffers considerable discomfort from his injury but rarely admits to it. A simple handshake can cause him problems and this is the reason that he greets people with a distinctive double-hand clasp. This habit, which seems so endearingly friendly, is in fact self-protective, to equalize the pressure that a vigorous handshake would impose on his injured back. At receptions where there are many people to greet, he will endure the routine unflinchingly; only *in extremis* will he avoid handshakes by wrapping a bandage around his uninjured right hand to give the impression of sprain. His first

secretary at Dome, Ethel Cairns, says she made it one
of her duties to assist Gallagher on and off with his
coat, since he sometimes finds the maneuver difficult,
and Dome's aviation staff ensure that their boss does
not have to squeeze awkwardly into one of the smaller
company planes. Gallagher, who is not given to fussing,
is unaware of this kindly interest in his comfort. Remarkably, his back injury has done little to impede his
interest in athletics. He runs several miles daily, skis
and plays golf well but infrequently, and in his sixties
took up wind-surfing at Lake Windermere in southeastern British Columbia, where he has a weekend cottage.

In 1948 Standard of New Jersey sent Gallagher on an
advanced-management course. "I believe Standard recognized in me an ability to meet with heads of government in various parts of the world and obtain
concessions, and then get a project going and complete
it in very rapid time, so they specifically sent me to
Harvard with the idea of further expanding my administrative capabilities. I'm not what you'd call a scientist. Geology interested me because you'd see and
work in remote parts of the world and help turn a desert
or jungle into something that's productive."

After Gallagher's stint at Harvard, Standard wanted
to appoint him the company's chief geologist for the Far
East. He had decided by this time, however, that he
wanted to return to Canada, and so he became assistant
to the production manager of the western divison of
Imperial Oil, Standard's subsidiary in Canada, at about
a third of the salary he would have got in the Far East.
After more than a decade of exposure to worldwide geology and the international oil business, he now turned
his attention to helping develop Canada's hydrocarbon
potential.

By this time Gallagher was already on his way to
becoming a wealthy man. He had managed his money
well during his overseas travels, living for the most
part on his expense account while purchasing Shell and
Standard Oil stock with his generous foreign salary.
"Jack is very shrewd. Jack always looks out for Number
One, you know," says an oilman who knows him well.

He joined Imperial at an important time: the year before, in February 1947, it had made its fabulous discovery at Leduc, eighteen miles southwest of Edmonton, after drilling a frustrating series of 133 dry holes. Leduc was the first oil discovery in Canada in eleven years and marked the start of Western Canada's modern oil industry. Imperial was now expanding its exploration and production departments rapidly, and Gallagher, a rising star, was obviously destined for great things. But he felt uncomfortable at Imperial and didn't stay long, remarking that he simply didn't fit into a large company. He had, of course, spent the previous decade and more working for the biggest of the multinationals, Standard Oil of New Jersey, but he had been his own boss in the deserts and jungles, far away from the bureaucracies and constraints of a head office. He was restless, energetic, and consumed with a bounding ambition. At that moment of dissatisfaction, he was approached by a group of investors with an intriguing proposition: how would he like to manage their affairs on the oil patch? Their vehicle was an unknown private company called Dome Exploration (Western) Ltd. and Jack Gallagher would be Employee Number One. It would be his show. The year was 1950, he was thirty-three, and the real adventure of his life was about to begin.

Chapter Four

THE HOUSE THAT JACK BUILT

When Jack Gallagher left the Standard* empire, in which he had been employed lucratively for more than a decade, he turned his back on a brilliant future with the world's largest multinational oil company. At Standard, bright young executives could step on a smooth career escalator and accumulate steady increments of pay, power and privilege as they rose towards corporate nirvana, comforted by the knowledge that they need never fret over the mortgage or the orthodontist's bills. Even if a vice-president's halo was not forthcoming, at least a generous pension would justly reward their labors as cogs in the service of Big Oil. By contrast, in 1950 Gallagher did not join Dome, he *became* Dome.

The new enterprise was the offspring of Dome Mines Ltd., a pioneer gold producer in the Porcupine district of Ontario that had been in business since 1910 (and is still in production more than seventy years later). The mining company faced a serious problem in the late 1940s: gold was frozen at an official price of $35 an ounce, yet production costs were rising steadily. Clearly it needed another string to its bow. Encouraged by the historic Leduc strike near Edmonton, the gold

---

*Gallagher worked for the Standard Oil Company of New Jersey (since renamed Exxon), not to be confused with either Standard of Ohio (Sohio) or Standard of California (Socal), which are separate companies, although all are progeny of John D. Rockefeller's old Standard Oil Trust. Exxon is the parent company of Imperial Oil.

miners decided they should spread their risks and put some money into the search for oil in Western Canada. So they formed a small subsidiary, Dome Exploration (Western) Ltd., and Jim McCrea, vice-president of Dome Mines, approached the restless Gallagher to manage it for them.

At first, Gallagher was so unsure of the company's prospects that the only person he hired for about two years was his secretary, Ethel Cairns. He had good reason to be cautious: of two hundred small oil companies started in the 1950s in the enthusiasm that followed the Imperial discovery at Leduc, Dome was the only one to survive intact. The fledgling company had a modest capitalization to engage in an expensive and risky business: $250,000 in equity and $7.7 million in debt, part of which was contributed by the tax-exempt Princeton, Harvard, and M.I.T. endowment funds and a number of private individuals, including John Loeb and Cliff Michel of the New York investment bankers Loeb, Rhoades and Company Inc., which had a major stake in Dome Mines, and Boston money-manager Bill Morton. Cliff Michel was at that time president of Dome Mines. U.S. tax law was generous in its treatment of investments in oil exploration companies, even outside the United States, and provided a pool of risk capital from which Dome would drink often.

The company's head office was in Toronto (where Dome still holds its annual meetings), and Gallagher set up a small Western office at 206 Alberta Block, Calgary, and spent $1,191 on furniture and equipment. It was not until 1958 that Dome adopted its present name and moved officially to Calgary. Gallagher's title at first was executive vice-president although it was he who ran the company, reporting authoritatively to six American directors and one Canadian director innocent of oil industry expertise; thus unencumbered, he was in the unusual position for a hired manager of not being second-guessed by his board. This fortunate circumstance endured for thirty years.

Gallagher was, in his own words, a person who loved to "wheel and deal". At Dome he was in his element and rarely stopped working. He drove himself hard, rushing from oil field to airport, where he would change his dirty coveralls for a business suit brought along by his wife Kay. Then he would board a Trans-Canada Air Lines North Star for a night flight to New York or Chicago for a board meeting. Arriving at about 10 a.m. he would shave at the airport and go straight into his meeting. Not bothering to check into a hotel, he would leave by plane the same day, flying back to Calgary at night, napping on the plane to be ready for another day's work. It was Gallagher who pored for hours over geological maps to select drilling sites, who took investors by the elbow and persuaded them to support his unknown enterprise, and who sweet-talked bankers, dazzled politicians, and inspired his employees.

Dome had a lucky start; at least, Gallagher, with engaging modesty, calls it luck. Its first two "wildcat" wells were producers: one at Drumheller revealed oil far removed from any known oil fields at the time, and, while not large, it has pumped steadily and profitably for over thirty years; a second "wildcat" turned out to be a gas producer in the big Provost gas field. Dome was able to clear its early debt primarily because of some step-out wells it drilled around the Redwater oil field developed by Imperial Oil. Under provincial regulations, Imperial had to give back to the Crown half of its acreage in each township within ninety days of starting production. Part of the land it surrendered was along the Saskatchewan River, where there was a thick gravel bed and where it could not get decent seismic information. Gallagher acquired this acreage and used a different seismic technique to discover the thickest pay in the Redwater field. Dome got its money back a couple of times over and then sold its interest for $8.7 million in March 1962, which enabled it to pay back its bank loans.

In about 1950, in the wake of Leduc, Calgary began to attract a growing technical fraternity of engineers and geologists. Gallagher helped found the Oilfield Technical Society, and there he met Charlie Dunkley, and evaluation engineer who worked for Chevron Standard and was secretary-treasurer of the group. At one time, Gallagher asked Dunkley if he would like to join Dome but he didn't press him, because, as he said years afterwards, he could never be sure when the company might go broke and he didn't want to lure anyone away from a secure job. Dunkley jumped a few months later, however, and worked with him closely for a quarter of a century, remarking shortly after he joined Gallagher that he never realized work could be such fun. They worked long hours, even on Saturdays and Sundays. Sometimes they had a quick lunch at the Petroleum Club, where gin rummy was popular in the downstairs card room. As the two were leaving one day, the work-aholic Gallagher remarked that the card players, lingering over their game, just didn't offer Dome any competition. (When Gallagher achieved celebrity status much later, the Petroleum Club began featuring on its lunch menu a Gallagher special—a turkey salad sandwich, tinned fruit, and cottage cheese.)

Dunkley believes one of the reasons Gallagher hired him was his own chronic pessimism, which served as a corrective counterpoint to Gallagher's unfailing optimism. Gallagher welcomed discussion and debate, and even as the company got bigger he still sought the opinion of everyone sitting around the table at management meetings.

When Dunkley finally retired in February 1979, he was a senior vice-president and a rich man, with a round-the-world trip for him and his wife as a parting gift from Dome. But Dunkley's fortune, like that of many long-time employees, rose and fell with Dome's stock market performance, and the 120,000 shares he still held in late 1982 were worth about $360,000, compared with a value of about $3 million a little more than a year before.

Ethel Cairns had worked for Gallagher briefly at Imperial, but, quite typically for him, he did not ask her to join his new company because that would have been "raiding". Instead, he advertised the position and gave her the job when she responded. She stayed with him for nineteen years but has mixed feelings about her former boss. "I always liked Jack, you could hardly help it. But I never did agree with his outlook on most things. When I met him it was not long after the war and he had spent the whole of the war in foreign countries. He was so nationalistic—but there was no way he would ever have gone into the army. I had three

brothers and my husband in the army. They all came
back with nothing and he came back with money to
burn." She also disliked his unprogressive attitudes towards women. "Jack is so set in his views that if two
people in the office were competing for a job and one
was a woman, there was just no choice; for sure the
man would get it. This was long before the women's
movement got steamed up."

Mrs. Cairns worked many Saturdays and Sundays
"if there was a flap on," and once even remained at her
desk until 11.30 p.m. on Christmas Eve to finish her
typing so that Gallagher's correspondence would be
ready for the first working day after the holiday. She
was never paid overtime; in fact her salary was far less
than that of other oil company secretaries in Calgary,
and she once calculated that in a six-week period she
had worked a full extra week in overtime. With this
work load, she complained to Gallagher that she should
not also have to make and serve coffee. With unhappiness, she recalls that he told her, "You can always
work somewhere else." She thought this a brutal response, since she had worked for him for about seven
years then, was over forty at the time, and faced little
prospect, she felt, of employment elsewhere. Gallagher
could be curiously insensitive and generous at the same
time: he offered Mrs. Cairns and her husband the use
of an apartment in Hawaii for a free winter vacation,
not realizing, however, that they could not afford the
air fare.

But his present secretary, Ardith Wagner, testifies
to his thoughtfulness. When her mother died and she
was feeling down, Gallagher arranged for her to fly to
California for a holiday with him and his wife in their
condominium at Pauma Valley, about fifty miles inland
from San Diego. "I realized afterwards that he must
have done it because my mother had just died and he
wanted me to have a break, but he did it so kindly. He
didn't say, 'You poor girl, you need to get away.' But
that must have been the reason." She noted that Gallagher, who once took a speed-reading course, works
his way through a voluminous correspondence and stays
on top of national and world events by constant reading,

and also discovered to her surprise one day that her boss could read back her Pitman shorthand notes.

The third person Gallagher hired to work for Dome was nineteen-year-old Maurice Strong, who, concerned that his youthful appearance would mask his abounding talent, grew a moustache to try to look ten years older. Despite his obvious cleverness, his grammar occasionally contained faults which Ethel Cairns attempted to correct, much to the chagrin of the impatient teenager. Years later, when he had left Dome to become an international businessman and public service hummingbird, flitting from continent to continent, he apparently forgot his irritation with the overzealous secretary: in Calgary for a public appearance and in need of someone to type a speech in a hurry, he sought help from his former critic. Mrs. Cairns laughed, "He would not have been able to find anyone else to read his handwriting." Strong, who has matured from a fresh-faced youth into an entrepreneur of fifty-four, enjoys a profitable career that has combined in a unique fashion the business world and international good works. His appointment by his Liberal friends in Ottawa to the chairmanship of the Canada Development Investment Corporation has promoted some observers to speculate that this may be the springboard to a long-delayed political career.

Dome grew slowly at first, Gallagher constantly working to raise money to explore for oil and natural gas in Western Canada and later in the western states. Contrary to its high-rolling image in recent years, Dome's early strategy was cautious, concentrating on taking "farm-ins" (that is, acquiring an interest in other companies' lands) close to existing production, and buying Crown reserves. Every cent was plowed back to strengthen the infant company, and Gallagher penny-pinched on expenses. Only once in a while would Dome gamble with what Gallagher called a "shoot-for-the-moon wildcat", for which risks he was usually able to bring in U.S. tax money through his Loeb Rhoades contacts.

Over the years Dome has found lots of oil, and its discovery rate has measured up well to others in the industry, but ironically, for a company that has attracted so much public clamor, Dome has never enjoyed an outstanding discovery such as West Pembina in Alberta or the Hibernia field in the Atlantic. The elusive "elephant", the giant find of which oilmen dream and on which corporate fortunes are made, has always eluded the company.

Dome went public in mid-1951 when it issued 500,000 shares at $11.22 in Canada and $10 in the U.S. (These prices reflected the difference in exchange rates.) It made a second public offering of another half-million shares in 1955, the last time it went to the public for equity financing. Since then its growth has been financed out of earnings or by borrowed money. It has never paid a dividend, and for thirty years, until its financial crisis, it reinvested over two and a quarter times its cash flow each year. When people complain that Dome does not pay taxes, by deferring its obligations until an uncertain future date, Gallagher says if the shareholders are willing to wait for their money, the taxman should display similar patience. (A former deputy minister of finance says, "I can't count the number of times Jack Gallagher has told me that his company doesn't pay dividends!")

Gallagher purchased his first shares in 1951 and has continued to invest in the company ever since, even buying shrewdly when the stock dipped to $3.80 in the early 1950s, although he laughs that he did not dare tell his wife of this purchase at the time. Gallagher has not drawn a salary since 1967; by November 1981 he and his immediate family owned 5,292,220 shares, which were worth $134.3 million at the company's stock market high in the spring of that year. By the end of 1982 their value had dropped to about $15.9 million and Gallagher owed the company $7.2 million which he had borrowed interest-free to buy stock.

As Gallagher became known around Calgary, he developed the reputation of an overly shrewd deal-maker, a man to be watched very carefully indeed. Another oilman once said of him, "He moves a ton of canaries in a half-ton truck by keeping them all flying." Dunkley tells the story of an out-of-town oilman who, seeking advice on how to tackle Gallagher, was warned, "Just be sure you don't lose your shirt." The man emerged from a long bargaining session still wearing his shirt and remarked on the fact proudly. "But where are your pants?" he was asked. Gallagher has also been called a riverboat gambler, which comparison he resents. "If I was a promoter, if I was a so-called riverboat gambler who always has the odds in his favor, we wouldn't be up in the Arctic," he has said.

Dome was to develop a reputation in the oil industry for its free-spending ways, but Gallagher prunes his own expense account rigorously. Once he had dinner at the Waldorf-Astoria hotel in New York with Strong and another Dome employee, Don Wolcott. Neither Strong nor Wolcott touched their bread rolls, and so their boss, intimidated neither by the splendor of the surroundings nor by the haughty expressions of the waiters, said, "If you don't want to eat those, we can send them back and save a couple of dollars." Even when Dome achieved the status of an energy giant, he refused to spend company money decorating the corporate offices with works of art (then very much in vogue in Calgary, where oil companies display impressive collections), saying that shareholders had invested their money to find oil, not buy pictures. This did not mean that the walls of Dome's executive offices were bare: he had no objection to displaying art works purchased by companies which Dome subsequently took over!

In the early days most of his business trips were either to Dome Mines in Toronto or to Loeb Rhoades in New York. Occasionally he would visit Winnipeg, where his family still lived, but he would religiously deduct from his expenses any charges related to family entertaining. His secretary, Mrs. Cairns, used to keep track of these expenses and she said later, "I always figured Jack was being more honest than he should have been. Other people would take their family out to dinner and charge it to expenses, but not Jack. There was no way he was ever going to let anyone say he was taking advantage of his position."

Gallagher has a sense of propriety that borders on the prudish: years later, when the company's Arctic operations were in full swing, he would not allow either his wife or Ardith Wagner, who had by this time become his secretary, to visit the Beaufort Sea on a company

jet, despite empty seats, because such a trip would have
smacked too much of a "junket". In 1982, when the
company was in deep financial trouble, he cancelled an
appearance at the annual Oilmen's Golf Tournament,
an ostentatious affair held alternately in the resorts of
Jasper and Banff, because he felt the chairman of a
company in difficulties should not be seen engaged in
such a frivolous pursuit. For years he had used the
Oilmen's tournament as an essential meeting place to
widen his network of business contacts.

One Christmas, Gallagher and his wife Kay and their
three boys went to Hawaii, and in his absence the staff,
then numbering about twenty-five, held a party in a
private dining room at the old Petroleum Club. Liquor
flowed too freely, and a couple of boisterous geologists
who had been discussing the Grey Cup were inspired
to demonstrate a few football plays. When their boss
returned and saw the bill for the broken dishes and
fixtures, he was mortified at the injury to the company's
reputation and swore they would never hold a party at
the Petroleum Club again. Gallagher doesn't smoke and
drinks very little, and it was understood that those who
could not carry their liquor would not go far within the
company. "I don't think Jack has ever bought anyone
a drink in his life," says a vice-president. So acutely
sensitive is he to the lingering smell of cigarettes that
smoking is banned on company jets if he is likely to
use the aircraft within the next few days. Gallagher
learned to eat sparingly while in the tropics; his lunch
often consists of bran muffins baked by Mrs. Gallagher
which he carries to work in his briefcase. These are
supplemented by fruit and carrots prepared by his office
staff.

Although he is chairman of Canada's third-largest
corporation, Gallagher abhors ostentation. Visiting the
office of Dome's new medical officer, Don Johnston, one
day, he remarked approvingly that there was no certificate on the wall and the doctor had not thought it
necessary to put the initials "M.D." on his name plate.
He even complains that he doesn't need the pretty flight
attendant to fuss over him on the company's business
jets, and the flight staff don't like to tell him she is
there in case he or the other senior executives are taken

ill during a flight. He once rebuked a visiting million-aire businessman because he was swaggering around Calgary in an oversized, chauffeur-driven Cadillac; Gallagher drives a modestly sized Cadillac Seville. He has lived in the same house on Britannia Drive for thirty years, and although he owns a condominium in California, it is mostly used by his wife as an escape from Calgary winters. On his infrequent visits to their holiday home he is amused to be introduced around as "Katie Gallagher's husband".

Gallagher married Kathleen Marjorie Stewart from Red Deer, Alberta, in August 1949, and they have three grown-up sons: James, a property developer, and Tom and Fred, both geologists. Being married to the work-aholic Gallagher cannot always have been easy, since the demands of his job often took precedence over family convenience. Planned weekend visits to the family cottage on Lake Windermere in southeastern British Columbia, about four hours' drive from Calgary, were often delayed repeatedly or postponed as Gallagher discovered he was unable to tear himself away from his office.

Gallagher's days usually begin, and often finish, with a jog along the Elbow River, but even while exercising, his mind is active, and sometimes he will dictate his thoughts by telephone into a recording machine in the Dome Tower the moment he returns home, the words interrupted by unusually heavy breathing. Once, after visiting one of the company's Arctic vessels which was being modified in Victoria, B.C., he decided that a group of overweight Dome sailors and marine engineers did not look fit enough, and so he had them jogging up and down the dockside as they waited for a taxi. At home, if he does not feel like jogging in the park, he uses a track in his back yard or exercises on a trampoline. In his late sixties he is still a keen downhill skier and cracked his ribs in 1982 when another skier ran into him. Typically, he scorns the easy slopes, telling com-panions they ought to be optimists.

When he went into hospital to have a calcium spur removed from his heel, he took phone calls right up to the minute he was wheeled away to the operating thea-ter. Later he required a second and related bout of sur-gery, after which he was fed fluids intravenously; even this did not stop him dictating memos and letters until the hospital staff began dimming the lights at night.

He has tried to run Dome much like the paternalistic owner of a corner grocery store: the supervisor of a pale-looking employee is asked if the man has had a recent medical checkup; office walls that are too glaring must be redecorated; light-covers in his dentist's office give a kindly illumination and so these must be installed in the Dome Tower; the cleaning staff is interrogated about the type of floor wax they are using; articles are photocopied and sent with notes to staff members they may interest; the company contributed $350 a year to annual athletic club dues (as long as drinks weren't served). Employees have been given free bus passes, a turkey each at Christmas, and an annual clothing gift certificate worth $160. The company held a traditional Western barbecue once a year in the fall, but this became difficult to organize as Dome swelled. One of the last ones was held in December 1981 at the Roundup Centre, part of the Calgary Stampede complex, where more than five thousand people ate steaks while being entertained by Tommy Hunter and several bands; since there weren't enough extra chairs and tables in Calgary to accommodate the entire Dome crowd, additional ones had to be brought in from Red Deer.

Gallagher rarely raises his voice, never loses his temper, and even under intense pressure usually conceals whatever irritation he may feel. That does not stop him from pulling rank firmly on subordinates when he feels it is necessary. Hearing a supervisor rudely reprimand a junior staff member, he approached the youngster with his charm turned up to full brilliance and said, "I don't think we have met. My name is Jack Gallagher." Then he turned coldly to the offending manager and inquired, "And who are you?" Despising nepotism, he will not allow the children of an employee to work for Dome, even during summer vacations, arguing that this is for the child's benefit, since he must be sure he has earned his own success and has not depended on parental influence. (A cynical senior executive says this is simple enough to circumvent: you get your friends in other oil companies to hire your children and you recruit their offspring as a *quid pro quo!*) Gallagher's rules also apply to the employment of spouses.

Remembering the frugality of his own field trips, Gallagher sometimes felt Dome, as it grew bigger, was pampering its workers. In particular he thought it an unncessary extravagance to supply fresh milk at the Rae Point drilling camp in the High Arctic. He had lived on powdered milk in the Andes and there was nothing wrong with it! His unpopular instructions were ignored, but on his next visit to Rae Point the crews saw to it that only powdered milk was on the mess-hall tables, a fact that Gallagher remarked upon with satisfaction. After lunch, ever curious, he asked what was in a shed and was told it contained "just some old machinery". However, an over-eager foreman, unaware of the secret contents, opened the door and Gallagher caught sight of a big stack of fresh milk cartons hidden there. He was not amused.

He is famous for his strong nationalist sentiments. One young executive was lectured earnestly about the need for pride in one's own country after the Dome chairman had discovered, by a chance remark, that the employee had bought a vacation cottage in the United States. The younger man took the admonition in good part, not thinking it tactful to remind Gallagher that *he* was the owner of a condominium in California!

Despite his famous smile, which has been photographed and commented upon *ad nauseam*, Gallagher does not possess much humor, nor does he ever kick up his heels at a party or celebration. "I think Jack's idea of a good time is to spend a quiet evening at home with a good book," one oilman, whose own idea of a good time is at variance with this practice, says disapprovingly. For a man endowed with social graces and political antennae for which an ambassador would eagerly give an arm, Gallagher is at heart surprisingly reserved, even shy. He is affable and approachable but not gregarious, and he does not take his own celebrity seriously. Nor does he seem to worry about personal security, an understandable fetish of the rich; at his secluded weekend cottage in British Columbia, his name is displayed at the end of the driveway for all the world to see. He endures interviews warily, like a patient

undergoing an unpleasant and mildly indecent medical
procedure, fearing that the probing will produce a diagnosis far from his liking. When he was evading the
news media with special determination at the height
of the company's 1982 financial crisis, one enterprising
Calgary TV reporter even took to jogging along Gallagher's accustomed exercise route, hoping, unsuccessfully as it happened, for a "chance" encounter. But
Gallagher would probably have appreciated the initiative, had he known; Mrs. Cairns says she is sure he
often engineered such "accidental" meetings with the
influential and powerful himself.

For a number of years, interviewers were themselves
interrogated first for possible bias by Bill Payne, the
company's public affairs director. Payne was hired to
manicure Dome's image after a disastrous bout of publicity during the first Beaufort Sea drilling season in
1976, when the CBC at Inuvik, which has strong con-
nections with native-interest groups, harassed the company continually. So well did Payne disclaim knowledge
of Dome's activities when he was required to stonewall,
as all public relations men must occasionally, that reporters sometimes wondered if he actually worked for
the same company they were inquiring about.

Payne, now a full-time politician, was another of that
strange breed of Dome workaholics. An MLA for the
affluent Calgary constituency of Fish Creek, he divided
his time between the Dome Tower and the provincial
capital, Edmonton, during the legislative sessions. He
began work at 7 a.m., fielding perhaps sixty phone calls
from reporters and shareholders, rushed to Calgary airport around noon, gobbled a packet of peanuts for lunch
on a Pacific Western Airlines commuter flight to Edmonton, and attended caucus sessions and then tedious
debates in the legislative chamber where Premier Peter
Lougheed, an iron disciplinarian, would not allow members of his overwhelming Conservative majority to tackle correspondence or read newspapers; he then skipped
out of the House around 9.30 p.m. for a flight back to
Calgary, where he tackled more Dome work at home
before getting to bed around midnight. His Dome salary
was supplemented by an MLA's pay of $29,962 (includ-
ing a $6,809 tax-free allowance). Such outstanding po-

litical service was rewarded with a cabinet seat as minister without portfolio after Lougheed's reelection in the fall of 1982; Payne was given the job of overseeing the image of Alberta's controversial Heritage Trust Fund and he resigned from Dome.

Dome's approach to publicity is certainly ambivalent: Bill Richards once sought to hire Earle Gray, author of *The Great Canadian Oil Patch*, to handle the company's public relations, and told him: "In the mind of the public Dome is a great, grey, vast unknown— and that's the way we want to keep it." In this, as in much else, Dome's approach was to prove flexible. It has courted publicity unashamedly when it has been to its advantage, and has been aggrieved when it has subsequently been unable to avoid it. There is scarcely a Rotary Club or chamber of commerce from Gaspé to Prince Rupert that has not been subjected to a slick Dome slide-show with futuristic ice-breaking tankers, transpolar navigation routes, and graphs showing highly optimistic projections of Arctic oil potential, or a financial analyst who has escaped the company's proselytizing zeal. Yet senior executives still appear mystified by persistent media attention, wishing naively that publicity could be turned off at will like a tap. When news about the company was uniformly black in the summer of 1982, a number of senior officers would have liked to shoot the messenger, and referred sadly to "hostile" press coverage when their own incautious actions came home to haunt them. Discussing the prospect of this book in June 1981, Gallagher said he did not really enjoy constant headlines. When reminded that these were inevitable, since his company was frequently involved in multi-million-dollar takeovers, he responded lightly, "I've got another one. I'm going to buy a shipyard tomorrow"—the purchase of Davie Shipbuilding Ltd. of Lauzon, Que.

Dome had come a very long way. The man who had spent less than $1,200 furnishing a rented office for a newly born wildcat oil company thirty years before was now paying $38.6 million to buy his own shipyard. And yet, by Dome standards in 1981, that seemed scarcely more than loose change.

Chapter Five

OTTAWA'S BLUE-EYED BOY

For years Jack Gallagher had said he planned to retire by the time he was fifty, so that he could work for $1 a year for United Nations agencies assisting the underdeveloped Third World. That he did not get around to this idealistic activity was because he found his work at Dome so seductively engrossing. His interests, however, far transcend those of the average businessman. He is an outspoken Canadian nationalist and has a remarkable empathy with those in public life; he also has the useful ability to clothe measures that are in Dome's crass commercial interest in the royal purple of public good.

Gallagher has thought much about the institution of government, and ways in which it could be improved, but there is an air of innocent impracticality about some of his plans. In July 1962 he told an audience of Rotarians in Calgary that the salaries of all cabinet ministers and Members of Parliament should be doubled and made tax-free, which brought the editorial response "Let's Not Go Overboard" from the *Calgary Herald*. He was concerned about the "near impossibility" of attracting top men into cabinet posts because of the uncertainty of tenure, the financial sacrifices involved, and the "almost inhuman work load". It was a serious weakness, in his view, that MPs were expected to have the combined talents of a vote-getter, a parliamentary debater, a policy-maker, and a director of a large government department. In this and later speeches Gallagher also proposed a number of radical structural changes in the federal government.

By 1974 Gallagher was thinking seriously of forming

a new political party to be called the Canada Party. He outlined his tentative ideas in an interview with Peter C. Newman in *Maclean's* magazine in September of that year. "The weakness of the Canadian political structure now is that we have one party whose power essentially emanates from Quebec, we have another party which owes its loyalty to the manufacturing interests of southern Ontario and a third national party whose allegiance is to organized labor. But we have no political movement that speaks for Canada's natural resources—the interests of agriculture, forestry products, fisheries, mining, oil and gas. What we need is a fourth party openly representing the natural resource sector and emphasizing individual initiative." Typically, he was wildly optimistic, predicting that by the 1978 election, after four years of careful preparation, the Canada Party could win between thirty and forty seats in "the resource constituencies" of Western Canada, the Mari and Northern Ontario. "In a minority parliamentary situation, we could then ask for and probably get three or four cabinet posts, such as agriculture, fisheries and energy, whatever party forms the government. This would give us a meaningful voice in policy decisions. By being members of a separate party working within a coalition, our MPs would have to be heard because they would be holding the balance of power." The *Calgary Herald* again took Gallagher to task and made the telling point that "resources don't vote." Despite Gallagher's enthusiasm and promised financial support, the Canada Party concept turned out to be a political dry hole.

But Jack Gallagher had established a reputation as a businessman who actually *thought* about the nature of government; a Canadian nationalist whose view was not directed with myopic exclusivity to the mercenary interests of his own company; a visionary whose latent desire to develop Canada's Arctic frontier happily coincided with a growing awareness in Ottawa of the country's vast unexplored riches. With such a handsomely engraved visiting card, Ottawa's doors were open to him whenever he came calling. And he did go calling, often.

* * *

In a moment of frustration in November 1981 when senior executives of Dome Petroleum were experiencing temporary difficulty in having Ottawa's budget changed for their benefit, one of the Calgary oilmen joked, "Perhaps we should be out on Parliament Hill carrying a picket sign." It is fascinating to speculate, in the circumstances, what message it might have carried: "Closing of Tax Loophole Unfair to Dome!"

Almost every night, the television news, a medium oriented more to entertainment than to information, shows demonstrations outside Parliament: native groups drumming and dancing, postal workers, prison guards, gays, feminists, human rights advocates, special-interest groups of a dozen persuasions, waving signs and banners, competing in their choice of bizarre costumes and loud banners for the satisfaction of seeing a twenty-second film clip on *The National*, and believing, usually erroneously, that they have thereby scored a major success in influencing the inhabitants of the corridors of power. Real influence, however, is exercised differently—not in strident confrontation but, as Dome does so well, quietly, persistently, persuasively, and professionally. The person who created Dome's political style, who *was* Dome's style, was Jack Gallagher.

Gallagher's apprenticeship as a political lobbyist was begun early in his career when he met General Jorge Ubico, an absolutist ruler, fourteen years in office, fiercely sensitive to criticism, whose arbitrary policies had aroused widespread unrest. Sometimes in Ottawa Gallagher must have felt a sense of *déjà vu*. That first encounter took place almost four decades ago in the ornate government palace in Guatemala City, when the young Gallagher extracted concessions (mostly the use of mules and canoes) from Guatemala's dictator. Such a minor-league try-out was just a start, and by the time he had graduated to Ottawa, Gallagher had honed and polished his early skills as a lobbyist until he was virtually without equal in Canada. By this time, however, he had set his sights on far more than mules and canoes!

For a decade and a half, at least, Gallagher worked the street in Ottawa with a determination and skill matched by no other oilman in Canada. Travelling airline economy at first and then in Dome's swelling fleet of corporate jets, Gallagher whisked from Calgary to Ottawa usually once a month, and often more frequently if occasion demanded. He attaches high priority to government relations; there is no mystique—to him it is just another aspect of management. Getting to know the whims of government, keeping track of bureaucratic movements, and interpreting the nuances of political alliances are as important a part of his job at the head of an expanding oil company as raising investment funds and moving oil rigs. From the beginning, Gallagher approached lobbying as methodically as a life insurance salesman zeroing in on potential clients: he drew up a list of "prospects"—ministers and senior mandarins he must buttonhole—and he tracked them down relentlessly, using his famous charm to soothe any irritation they might feel at his persistence. A former deputy minister recalls that his secretary, a formidable office guardian with many years' experience in repelling unwanted visitors, invariably melted before the Gallagher smile.

Alastair Gillespie, who was Minister of Energy, Mines and Resources from September 1975 to June 1979, remembers Gallagher's persistence with amusement. "He learned the technique of gaining access by learning the habits of ministers. He would come down to Ottawa and I don't think he would make a decision about when he was going to go back to Calgary. If he had come down to see some people and they were on his list, he would stay until he had done so. He was very purposeful about it. I can remember on one occasion coming into my office at 8.30 in the morning and who was sitting in a chair in my outer office but Jack Gallagher. He didn't have an appointment. How does a minister coming in at 8.30 a.m. say 'no' to a man who says he just wants ten minutes of your time? He would be very businesslike and he always had a reason for wanting to see you and you always felt that what he had to say was worth listening to—and that, of course, was part of his magic. People didn't turn him down, because they knew he had something interesting to talk about. He probably understood better than almost any other businessman how the government worked and what the government's objectives were. He always had a very positive suggestion: 'Here is a way of gaining your objectives.'"

Gordon Harrison, who came from a senior position at Mobil in 1974 to head Dome's Canadian Marine Drilling Ltd., found working with Gallagher an eye-opener. "He has sensitivity not only to the technical aspects of the problem but also the politics. I have never seen Jack lobby in the normal sense. I have never seen him take people out to dinner and that sort of stuff. He is a very austere guy. The idea of sitting down and chitchatting and having dinners with people just would never appeal to him. He used to jam visits with so many people into a day that it was really an efficient trip. He would line all these appointments up—it was all bang, bang, bang from one guy to the next, ministers and deputies. I have seen him take a half-hour or a shorter period with a minister and try to cram as much information of a factual nature into that period as he can. Jack is on a first-name basis with a lot of them, but I have never seen any close friendships at work. For years we never employed a professional lobbyist; we used to go down ourselves. I think this was a breath of fresh air in Ottawa because we knew what we were talking about and came with the authority to make commitments."

Jack Gallagher made an astute decision by joining the board of the Canada Development Corporation at a time when the federal government was finding it difficult to attract competent businessmen to serve on the CDC. That made him a number of friends in Ottawa, including a highly placed new one, fellow CDC director Tom Shoyama, Deputy Minister of Energy, Mines and Resources, who subsequently became Deputy Minister of Finance. When Gallagher began to lobby for special incentives to help pay the cost of Beaufort drilling, he was to find a sympathetic listener in Shoyama, who saw Gallagher as a helpful and enthusiastic Canadian nationalist whose company was willing to pursue the national interest in ways that contrasted sharply with the myopic self-interest of such multinationals as Imperial Oil. Shoyama also regarded Gallagher as a valuable window on the energy industry, conveying to federal bureaucrats his view of the Alberta oil industry and those of the Alberta government. Shoyama found it useful to ask, "What do you think of this, Jack?" And, since the Calgary oilman was a frequent visitor to New York, "How are the Americans viewing the situation in Canada, Jack?" In return, Gallagher quizzed the mandarins at EMR in Ottawa on their views of OPEC or intelligence emanating from the International Energy Agency. In short, Gallagher was party to a mutually beneficial two-way conduit of information enjoyed by few of his peers in the Canadian oil industry.

However, Gallagher's reputed ease of access to Prime Minister Trudeau is certainly exaggerated. Jim Coutts, who was Trudeau's principal private secretary says he can remember only one occasion in over six years when Gallagher had a private meeting with the Prime Minister, and it was to discuss a subject (unspecified) other than Dome. That was not to say that Gallagher and Bill Richards, who later took on the greater burden of Dome's activities in Ottawa, did not occasionally meet the Prime Minister at crowded social gatherings. Gallagher also escorted Trudeau on a tour of the company's Arctic operations in 1981.

Gallagher's technique was quiet and simple—and devastatingly effective. He always had something to offer the policy-makers: a tidbit of intelligence, an idea, a policy initiative that would be helpful to those in the seats of power. Almost incidentally, it would also be helpful to Dome. He never stalked in angrily and told them their plans were hogwash—in the manner, for example, of one abrasive Calgary oilman who needlessly infuriated Liberal MPs at a Commons committee hearing by insisting on referring to the National Energy *Pogrom*. Rather, Gallagher made helpful suggestions—a slight change in direction here, a modification there, a tax concession perhaps—and assured them that Dome, ultra-nationalistic Dome with big red maple leaves plastered on its ships, would be only too happy to lead the charge in the direction that Ottawa wanted the oil industry to run.

Eventually Gallagher became so accommodating, so willing to run at the government's behest—and his company benefitted so greatly from its connection with the government—that Dome came to be known as Ottawa's "chosen instrument" on the oil patch. One civil servant who has been intimately involved in Dome's affairs says the tag is unfair. "They were not the chosen instrument in the sense that anybody chose them—they volunteered!" he insists.

A myth has emerged in recent years about the nature of Dome's relationship with Ottawa: the vague perception of sinister oilmen cleverly manipulating the custodians of the public purse for their own venal benefit. Dome's activities, however, were straightforward, open, and brisk; certainly there is no hint of the corrupt influence that enlivens political fiction—sealed envelopes stuffed with hundred-dollar bills being passed across ministerial desks. There isn't even much evidence of entertaining. As for the company's political contributions, they were mere peanuts—and more or less evenly distributed. In 1981 Dome Petroleum donated $1,800 to the Liberal Party and $768.90 to the Tories. That same year, its sister company, Dome Mines, gave $1,000 to the Liberals and an equal amount to the Tories. TransCanada PipeLines, which Dome controlled, gave $1,800 to the Tories and nothing to the Liberals.

Dome did not always succeed in its efforts to extract concessions out of Ottawa. It foundered badly when it sought to get the federal government to pay the lion's share of the construction of a Class 10 Arctic icebreaker in the late 1970s. Alastair Gillespie recalls that Gallagher very nearly did get his way. "He was too greedy, quite frankly. My impression was that he wanted the government to pay ninety-five per cent of it but wasn't prepared to let the government use it for more than a fraction of the year, and that fraction of the year was the least important time. He was dragged screaming and kicking into allowing even that joint use. And then it became apparent it was going to be stationed in the part of the Arctic [the Beaufort] where the Coast Guard could not have used it. I think if he had bent a little bit on that, and if he had also accorded to PetroCanada some use of it in the Arctic, then he might have carried the day. He was just being a little too greedy. But conceptually, think how exciting it was: this Arctic Marine

Locomotive that would charge through ten feet of ice at three knots. It was a very well-sold idea."

Gordon Harrison gives Dome's version of the aborted super-icebreaker. "We thought there were many factors that justified its construction. One was opening up the Arctic. We were trying to get assistance from the government because we were looking at the future, rather than the present. We were saying, 'If you can move around in the Arctic year round, the resources there will be more available.' But Ottawa just didn't find a justification in the economic climate at the time to support another icebreaker. I suppose it is a little like the building of the CPR. Do you build it because there is enough business to support it or do you build the railway because it will encourage the business? We thought Canada should have an equivalent of the CPR through the Northwest Passage. Eventually the government decided that since it was to be used for [commercial] development, we should get PetroCanada to support it, and they didn't. I think PetroCanada simply didn't want to put their funds into such an enterprise if they could not see a minimum justification for it."

In the wake of the OPEC oil embargo during the winter of 1973-4, the federal government suddenly became nervous about Canada's future oil self-sufficiency and took a keen interest in obtaining an inventory of the country's hydrocarbon resource potential. By 1975 Canada had become a net importer of oil, a reversal of the position just two years earlier, and the National Energy Board was estimating that foreign oil might amount to about half our total needs by 1985. So far the potential of Canada's frontiers didn't appear to be particularly impressive: the oil industry's efforts off the east coast had diminished to virtually nothing; the Panarctic consortium working in the Arctic Islands had discovered large quantities of natural gas but not much oil; and the Mackenzie Delta region, which had earlier displayed great promise, was also proving to be gas-prone. Canada, it seemed, had one last card to play and it could be a trump: the Beaufort Sea. No one knew a great deal about the Beaufort—although it was geologically appealing. To find out what hydrocarbon riches it might

contain, wells would have to be drilled there and costs
were certain to be extraordinarily high, partly because
of the added environmental constraints that would be
imposed on any company operating in the exceptionally
fragile Arctic.

Jack Gallagher, a constant visitor to Ottawa, knew
the thrust of government thinking, which by then dovetailed beautifully with his personal dream of Beaufort
exploration. Private ambition nestled comfortably
alongside the National Interest. Ottawa wanted to
know—indeed was desperately anxious to learn—what
riches the Beaufort might contain, and obliging Jack
Gallagher would find out for them. At a price.

As early as 1974, Gallagher had obtained a letter
from the then Minister of Northern Affairs, Jean
Chrétien, which said that if the company completed the
necessary environmental studies, the government would
grant it permission to drill exploratory wells in the
Beaufort. On the strength of that letter alone, Gallagher had committed Dome to buy two drill ships and
various support vessels at a cost of well over $130 mil-
lion. Few executives put such trust in the words of politicians. Gordon Harrison says Chrétien's letter gave a
signal to the entire Ottawa bureaucracy that the issue
of Beaufort drilling had to be dealt with. Without it,
he suggests, the civil servants would simply have found
ways of dragging their feet.

For a time, however, getting drilling permits hung
in the balance. Ewan Cotterill,* who was Assistant
Deputy Minister of Northern Development in 1976, recalls taking the initial drilling authority through cabinet. "There was quite a real possibility Dome was not
going to get the drilling authority. There was considerable opposition at that time. The general feeling was
that there had been an awful lot of environmental studies carried out, but there had never been a review or
assessment of the whole concept of drilling in the Beaufort Sea. When the time came to actually issue the
drilling authority, there were environmental groups,

---

*Cotterill has since left the federal civil service and now works for
  Dome.

native groups, and a loud public opposition to drilling at all, and it was by no means certain that they were going to get it."

By that time Judd Buchanan had become Minister of Indian Affairs and Northern Development, and he and Alastair Gillespie were the standard-bearers for Dome in the debates within cabinet. "There is no question that it was a hotly debated issue," says Buchanan. "There is always a tendency in government—more so at the official level than at the political level—to hide behind a negative response, and then you can't be criticized for doing something. As a politician your neck is out there for sure; it is the nature of the beast. If you don't want to risk that, you should not be in the business. But we imposed some fairly stringent conditions; I did not want to be the minister that had authorized or supported an action that would wind up doing considerable harm to the environment of the North."

Dome went to great efforts to convince Buchanan—and others concerned about the hazards of Arctic drilling—that they could cope if anything happened. As Gillespie remembered later, "There were many in cabinet who identified with the environmental approach. Warren Allmand took over the northern development portfolio after Buchanan, and he, of course, was very much in the environmental wing by instinct; but by then the track record of Dome was sufficiently impressive. They had shown that they knew what they were doing, and they had engaged and were training a very large number of Inuit people."

Nevertheless, the cabinet did delay the granting of Dome's drilling permission for several weeks at the start of its second season in 1977 because of environmental worries. That spring, a dramatic blowout at the Ekofisk oil field in the North Sea had captured headlines for weeks and stimulated renewed concern about the likely effects of a blowout under the Arctic ice. Dome was operating, in Harrison's words, in a goldfish bowl. "We got the first approvals by going to the cabinet, but then all subsequent changes of even a technical nature in our programs also had to go back to the cabinet. So if, from time to time during the season, anything fairly major came up, we had to go back to the cabinet ministers and all the deputy ministers, and in effect brief them so that the cabinet had an understanding of the problem. And of course one of the difficulties is that the ministers keep changing. We really had to get a consensus in the cabinet. And you don't get that necessarily by going to a majority of the cabinet people; you get it by going to the key people who are involved in any one decision."

Another factor which weighed heavily in Ottawa was the issue of Canadian sovereignty in the Arctic, which had been challenged in 1969 when the U.S. supertanker *Manhattan* crashed and battered her way from the Atlantic through the Northwest Passage to Prudhoe Bay, Alaska, where oil had been discovered the previous year. Humble Oil, which operated the *Manhattan*, at one time planned a fleet of thirty icebreaking supertankers to carry Alaskan crude oil 4,500 miles from the North Slope of Alaska to the U.S. east coast. There was a clear challenge to Canadian sovereignty when the U.S. government announced it would be sending a Coast Guard vessel to accompany the *Manhattan* and did not respond to an informal hint from Ottawa that the State Department should apply for permission. Eventually the sovereignty threat, seen as so important at the time, cooled considerably—partly because the *Manhattan* had to be assisted through the Northwest Passage by a *Canadian* icebreaker when she got stuck in ice, and partly because of the subsequent enactment of the Arctic Waters Pollution Prevention Act, which the Americans have not challenged. The sovereignty issue, however, contributed to the political climate in which Dome was given permission to drill in the Beaufort. If the Beaufort was *occupied* by Canadians, the cabinet felt, there would be one more argument against legal challenges to Canadian jurisdiction in future.

But Gallagher wasn't ready to establish Canada's sovereignty, or explore her northern resources, without a return. He argued that Dome could not carry the ball for the government in the Beaufort without generous tax incentives, and the result of his persuasiveness was a tax measure that became known variously as "super-depletion" or "The Gallagher Amendment", although its formal title was "The Frontier Exploration Allowance". Gallagher proudly acknowledges his role in persuading the government to grant super-depletion. It was his initiative and he was completely alone, he says. "I met Tommy Shoyama a few times, and then met with Don Macdonald, who was the Minister of Finance, and it was fashioned in his office. I suggested that was the only way they could get the frontiers going. I remember they were talking about making the super-depletion only applicable north of the 70th parallel, which would have benefited companies in the Arctic Islands and us [in the Beaufort] but not helped anybody else much." The 70th parallel passes just south of Dome's Beaufort acreage. "It would have benefited Imperial a little but not very much. My reaction was, 'Why not make it all of Canada? What have you got to lose? And then it doesn't look like you're zeroing in on any one place.'"

Donald Macdonald did not take much persuading, says Gallagher. "He's a very good listener, and very knowledgeable, because he'd been Minister of Energy and he's been in the Arctic. I found Don Macdonald grew tremendously during this period that I'd known him, when he was the Minister of Energy and subsequently Minister of Finance. And I think, personally, he would make an excellent prime minister because he's courageous and yet he has an excellent background both in government and out of government."

Macdonald introduced the Frontier Exploration Allowance in his budget on March 31, 1977. Super-depletion was, in fact, a generous icing that Ottawa laid over existing taxation incentives to stimulate frontier .esource exploration. As a result of the allowance, Dome was to live very comfortably on public funds— or to be more precise, funds that would have been public had they not been diverted before Revenue Canada laid hands on them—for three years. The incentives would apply only to wells costing more than $5 million—and it just so happened that there was only one company in Canada drilling wells that expensive at that point: Dome! In August 1979, the *Calgary Herald* calculated that of the $150 million spent in the Beaufort Sea in 1978, Canadian taxpayers (through taxes forgone) had contributed anywhere from $130 million to $140 million. The full implications of super-depletion were not at first appreciated by the senior civil servants in Ottawa charged with safeguarding the taxation system, and when the cleverness with which Dome had exploited the "Gallagher Amendment" became clear, more than one senior mandarin was infuriated. Dome conjured up a maneuver described as "tax sharing", which Bill Richards called "selling super-depletion". A wealthy individual, usually a high-income professional such as a doctor, a dentist, or a lawyer, who could afford to invest $10,000 in one of Dome's drilling funds, was able to write off $16,666 against his taxable income, *but* if he happened to have income from oil and gas or mining companies he could write off a grand total of $20,000. Even if Dome found not one drop of oil, certain investors would come out ahead simply by Dome's exceptionally adroit navigation of the tax system.

Through this device Dome raised about $170 million for frontier exploration during the time super-depletion was in effect. In addition to the private investors who put money into its drilling funds, Dome also was able to raise substantial amounts from other oil companies, who helped pay the cost of running its Beaufort Sea drill ships, courtesy of the remarkable tax concessions.

Gallagher makes a strong arugment that super-depletion was not simply a thoughtless giveaway of public funds. The country has benefited enormously, he says, for without the tax concession the giant Hibernia field in the Atlantic, discovered by Chevron Standard Ltd. in 1979, would never have been drilled, never mind Dome's wells in the Beaufort. When he made this argument to Pierre Trudeau, the Prime Minister responded that Chevron Standard, a U.S.-controlled company, had acquired its valuable Hibernia interests practically without cost to itself, and at the taxpayers' expense. Gallagher patiently pointed out that it was in Canada's interest that a whole new hydrocarbon basin had been discovered and Ottawa would be the net beneficiary eventually through the ripple effect of increased economic activity. To which Trudeau, unimpressed, replied: "Well, Jack, all I heard is that they got it for nothing."

* * *

By 1980 Jack Gallagher's visits to Ottawa had become somewhat less frequent, as his thoughts turned more to retirement; Bill Richards, who assumed the dominant role in running the company, also became Dome's main spear-carrier in Ottawa. It was he, as much as Gallagher, who skillfully turned a potential disaster to advantage when a new energy minister with some alarming ideas arrived on the scene.

When Marc Lalonde switched portfolios from justice to energy, mines and resources on March 3, 1980, he had scarcely settled behind his desk before Richards, his first visitor from the oil industry, dropped by to get acquainted. Very shortly afterwards, Gallagher, whom Lalonde knew by reputation as a bit of a snake-oil salesman, also called for a chat. In the months that followed, both the chairman and the president of Dome became persistent visitors to Lalonde's office, inhaling the political breezes with a diligence unmatched by any other company in Canada. The minister's former assistant, Mike Phelps, says, "We saw more of them than any other company, including [Crown corporation] Petro-Canada."

The men from Calgary cannot have enjoyed the new political atmosphere in the energy minister's office, for Marc Lalonde is an unswerving economic nationalist who, at the head of Ottawa's most powerful ministry, was determined to put his political stamp on the country with the National Energy Program, an inept and misguided policy that was anathema to both his Dome visitors. Although they found many aspects of the NEP distasteful, they accepted the change in direction, and adopted the old business cliché "If you can't beat 'em, join 'em." The newly elected Liberal government might, in Dome's view, be wrong-headed in much of its new energy policy, but it clearly had not only the ability but also the unshakable determination to fulfill its mandate to "Canadianize" the oil industry, and would do so no matter what they thought. Dome's creation of Dome Canada to conduct its exploration activities, as already related in detail earlier, sealed the company's close identification with Ottawa, to the horror of much of the oil industry.

By endorsing the NEP at a time when Marc Lalonde

was the subject of almost universal hate in the oil industry, Dome offered practically the only political comfort the strong-willed and beleaguered minister received from any source for many months. Lalonde, it is said, feels a keen sense of gratitude to Dome for coming to his aid at a difficult time in his career (although, surprisingly, he does not feel he has been injured politically by Dome's subsequent problems). Politically Lalonde incurred a huge debt to Dome. The train of events he set in motion with the NEP encouraged Dome to embark on—and the reckless banks to finance—the grand and foolish adventure that precipitated its downfall. It is obvious that without Lalonde's NEP, Dome would never have contemplated the HBOG takeover. Unfortunately for Dome, political debts have no collateral, their value being dependent on volatile public moods, the expediency of the moment, and the shifting of personalities.

The intriguing question is whether the favor Dome did Lalonde has been reciprocated. Did it have any bearing on subsequent actions by Ottawa in helping avert the company's bankruptcy? The likelihood is that it did not; Lalonde may have felt a personal political debt to Dome, but by the time the "bail-out" was complete, a new energy minister, Jean Chrétien, was dictating Ottawa's terms. Chrétien is in no sense beholden to Dome—to which the severity of the bail-out deal bears eloquent witness. In the end, when Dome *really* needed help from Ottawa, to an extent even greater than it had required previously, it received not a warm handshake but a back-handed slap to the face.

Chapter Six

THE NEW BREED

"When looking at Dome, there is always the danger of falling into the trap of the blind men examining different parts of an elephant. Each reaches a conclusion substantially different from the others about the true nature of an elephant, depending upon the particular part that he feels."

> —Bill Richards upon the listing of Dome Petroleum on the London Stock Exchange, June 2, 1980

For almost a decade, until it ran headlong into financial disaster in 1982, Dome was the shining Canadian success story, a home-grown mover and shaker blessed with the same Midas touch that elevated such high-flying American corporations as Xerox, Polaroid, and Texas Instruments into corporate superstardom. Dome was sleekly coiffured and well-tailored—but what was frequently overlooked was its well-worn and muddy work boots. For all its later glamor, Dome had from its early days shrewdly developed a very pedestrian and incredibly lucrative business processing and transporting natural gas liquids (NGL). This is an activity so uniquely unexciting, especially when compared to drilling for oil among the polar bears, that it is rarely mentioned in discussions even in the financial press. But not long after Dome's birth, Jack Gallagher had decided that the life of a wildcat oil company was too precarious to guarantee long-term growth.

"We recognized that unless you're extra lucky, you

cannot live on exploration results alone, because you can go for five or ten years with a long dry spell. You need to have quite a large and growing cash flow in order to stay with the majors in frontier plays." (Evidence of this vulnerability was the company's expensive 11,000-foot Buckinghorse test well, one of the earliest and deepest in British Columbia, which turned out to be dry.)

When Gallagher decided he wanted to establish a solid bread-and-butter business to balance the company's heavy exploration spending, he recruited Don Wolcott, an engineer with considerable natural gas experience then working for Gulf, who is, by common consent, one of the great innovators of the Calgary oil patch. Wolcott sheds interesting light on the decision-making methods of a multinational oil giant. When he was with Gulf Canada, his company proposed a $3 million expansion project in which Wolcott was involved, and so he and two other executives flew to the parent company's headquarters in Pittsburgh to state their case to the board of directors. Supplicants were allowed one minute for each million dollars they wanted to spend. Wolcott and his companions waited from 8 a.m. to 3 p.m., along with other minor executive cogs, in a corridor outside the Gulf boardroom, where a shoeshine man plied his trade. To pass the time Wolcott decided to have his shoes polished, but this operation was only half complete when his name was called. He made his pitch to "those big, tough guys" for $3 million with one shoe gleaming and the other dull. He got his money, though and the investment paid for itself in twelve months.

Wolcott was to find the atmosphere at Dome quite different. In 1957, the year he joined the company, it made a profit of only $204,131, and he moved into an office smaller than the one occupied by his former secretary at Gulf. But Wolcott was attracted to the smaller outfit because he wanted to run his own show; no longer would he be required to cool his heels in a corridor in Pittsburgh. He was employed initally by Provo Gas Producers Ltd., which Dome controlled with a thirty-seven per cent interest and eventually acquired, and his first job was to bring the Provost gas field in Alberta into production, which he did in August 1957. He stayed with Dome for twenty years, overseeing the development of the natural gas system and working miracles for Dome's cash flow, and then he quit, partly because he was miffed at being passed over for the presidency, which Gallagher gave to Bill Richards.

During the development of the Provost gas field, about 350 miles east of Calgary, a bizarre accident occurred. Dome hired a contractor to build three company houses for the personnel who would be manning the Consort gas plant. The company then drilled for water in this richly gas-prone area. Ed Tovell, who was Dome's operations manager until his retirement in 1975, recalls: "We turned on the water system and the contractor went into the bathroom and lit a cigarette. There must have been a gas accumulation in the water tank. He was blown clean out of the can with his trousers down. His posterior was brilliant red and he had to fly all the way to Calgary on his hands and knees because he could not sit down. It was two weeks before he could get back to work to finish the housing contract, although he was not too badly burned."

The same month that the first deliveries were made from the Provost field, Dome and Provo formed Steelman Gas Ltd., which was granted a permit to build facilities worth $10.7 million for gathering and processing gas previously burned off as waste in oil-well flares in the Steelman area of southeastern Saskatchewan. Within a year of its completion, the Steelman plant was already being expanded to process further volumes of gas, and extract propane, butanes, and natural gasoline from gas that in more wasteful times had been regarded as an irritating by-product of oil production.

In 1961 Dome pulled off an industry coup (although there was some scoffing about the impracticality of the scheme at the time) by building a facility that would extract liquids such as propane, butanes, and pentanes from natural gas as it travelled from the oil fields to customers by pipelines that linked up at Edmonton. Producers in the field were already extracting some of the liquids from the natural gas so that it could flow through the pipelines, but Dome hit upon the idea of

building one large central installation, a "straddle"
plant, which, as its name implies, sat astride the main
natural gas pipeline arteries of Northwestern Utilities
Ltd. and removed liquids still remaining in the gas
which the producers with their small field plants could
not extract profitably.

It was a clever decision, and although Dome had to
struggle at first to find markets for its production, it
formed the nucleus of an enormous integrated natural
gas liquids system that Dome expanded progressively
through a series of increasingly complex corporate alliances until it became the largest natural gas liquids
producer and marketer in Canada.

In 1970 Dome completed a further straddle plant,
owned equally with Amoco Canada Ltd., at Cochrane,
Alberta, which sat astride the Alberta and Southern
pipeline transporting gas to California. The key to
Dome's success in the NGL business was its ability to
accumulate sufficient volumes at Edmonton so that it
could send them to Sarnia, Ontario, in big "batches"
along the 1,800-mile oil pipeline operated by Interprovincial Pipe Line Ltd. The transportation savings
were dramatic: whereas it cost 10 cents a gallon to send
propane by rail, Dome could move the product by pipeline for as little as 1½ cents a gallon. To secure this
saving, though, it had to invest in costly pipeline and
storage facilities. Dome added further to its supply
in 1972 with the addition of a third straddle plant at
Empress, Alberta, which it owned together with Pan-
Canadian. This incorporated some expensive new technology for the efficient extraction of ethane, a risky
gamble since Dome had no immediate market for the
material. Eventually its search for a buyer led to an
argument with Columbia Gas Systems, an American
company that would buy the ethane and other NGLs for
a synthetic natural gas plant at Green Springs, Ohio.

Dome's natural gas empire has become a colossal
corporate octopus with tentacles extending from the gas
fields of Western Canada through a large and sophisticated processing and pipeline system to the industrial
heartland of eastern Canada and the eastern U.S. The
system comprises a series of neatly interlocking operations, and while Dome, as a rule, does not own more

than half of any individual piece of the jigsaw, the company runs almost the entire system on behalf of itself and its partners. In addition to the sale of hydrocarbons from its own walls, it is a trader, buying and selling the production of other companies.

With its natural gas business growing rapidly, there was an obvious logic to Dome's next move: the control of one of the world's biggest natural gas transmission companies—the huge TransCanada PipeLines Ltd., whose network of 5,800 miles of underground pipe moved up to three billion cubic feet of Western Canadian gas daily into eastern markets.

TCPL was controlled through a thirteen per cent interest by CP Investments, a subsidiary of the Canadian Pacific Railway, which announced in 1978 that its shareholding was for sale. The prospect of a shift in control of TCPL created great interest among oil and gas producers; the company was good at its job technically but an unadventurous marketer (witness the 200 billion cubic feet of gas it had contracted to buy from producers but could not sell). There was speculation that the Alberta provincial government might pick up Canadian Pacific's 5.1 million shares, but if it had any such ideas these were publicly discouraged by federal Energy Minister Alastair Gillespie. When Dome stepped in, there was a sense of enthusiasm within an industry burdened by an alarming oversupply of gas. If anyone could stir up the sluggish TCPL and start selling the product, it would be the aggressive Dome management. Not content merely with acquiring the CPR interest, Dome went on to make further open-market purchases of TCPL shares until it obtained forty-nine per cent control.

Dome revitalized TCPL. Gallagher, Richards, and John Beddome, a Dome senior vice-president, replaced three CPR directors, and Gallagher set about finding his own president and chief executive officer. He first approached Robert Bandeen, the president of Canadian National Railway Co., whom he later tried to recruit as Dome vice-chairman, but Bandeen wasn't interested in either offer and early in 1982 he became president of Crown Life Insurance Co. Gallagher then recruited Radcliffe Latimer, president of CN Rail (the largest and most profitable operating division of the CN empire) on December 1, 1979. Under Dome's stimulating influence, TCPL began a number of ambitious schemes to market more gas.

Dome's strategy was to get maximum control over its numerous acquisitions for minimum cost. Once within Dome's familial embrace, TCPL began making a series of kindly purchases that obligingly eased its parent's financial obligations without denying it an authoritative voice in family affairs. The first acquisition by TCPL, which established by contractual commitment the pattern for others that followed, was the purchase for $315 million of a 12.5 per cent interest in Dome's onshore oil properties in 1979. (The purchase actually was made from a subsidiary of Dow Chemical, not exactly a relative, but a close family friend of Dome, to whom they had previously been sold.) Later TCPL acquired 12.5 per cent of the oil properties that Dome bought from Kaiser Resources.

It didn't always work smoothly, though. When Dome acquired HBOG, it wanted to sell TCPL 12.5 per cent of HBOG's *Canadian* assets only. However, Radcliffe Latimer, the head of TCPL, insisted that the package must include part of HBOG's oil fields in Indonesia and Australia, and Dome argued for some time that this wasn't part of the understanding. Eventually TCPL got its way and paid $560 million for its slice of Dome's latest acquisition. Dome had outsmarted itself, however. Subsequently, when Dome, desperately needing cash, tried to arrange a quick sale of the Indonesian oil fields, TCPL dug its heels in and frustrated the deal for a while by refusing to part with its share.

The decision to move into the natural gas business was only one of the astute moves the young company made. For the first twenty-five years of its life Dome had grown steadily and undramatically, a highly respected player on the oil patch, showing the correct economic instincts, marketing savvy, and technical innovation to create an impression of solid accomplishment among its peers. Jack Gallagher had built a strong team of managers and was himself well regarded—indeed, among some of the younger oilmen, almost idolized—as a geologist with the right insights and a good story to tell investors who furnished the lifeblood of a growing oil company. Over the years Dome accumulated a land position of growing importance in Western Canada, the Arctic, and the United States.

Then suddenly, like a tornado appearing without warning, in the late 1970s Dome unleashed a frenzied burst of activity probably without equal in Canadian economic history. For a while it seemed that the company was being driven by an irresistible urge to become the dominant force in the Canadian oil industry overnight, with just a few throws of the dice. A number of old-timers who had been with the company since its early days fretted at the activities of a cadre of newcomers, corporate planners, and takeover specialists, who had the ear of Bill Richards, Dome's new president. The new men were sharp as a pin, lean over-achievers in three-piece suits, who seemed to be far more intimate with the subtleties of corporation law and the arcane workings of the Canadian taxation system than the routine business of land acquisition and drilling, producing, and transporting oil and gas.

One of them headed Dome's acquisitions department, which sewed together deals in the hundreds of millions and then in the billions. Wayne McGrath came to the company in the fall of 1976 from an unlikely background; he had just spent two years pumping gas and trying to set up a towing business in a small town in the interior of British Columbia. His business credentials were a little more orthodox than that would suggest, however. After graduating from the University of Saskatchewan, he worked in operations research for Federated Cooperatives in Saskatoon and then for a group of management consultants in Calgary, at which job he travelled so extensively that he logged more miles one year on Air Canada than his brother who was a pilot for the airline. McGrath's two years as a small-town businessman were his way of "copping out" of the hectic business world for a more leisurely change of pace. When he got back into the corporate world, through a casual phone call to a friend at Dome while killing time between flights at Calgary airport, he was to return to the frenetic hustle and bustle of business life

with an intensity that few executives have encountered
and most would dread! First as part of the business
development group, then as head of an elite acquisitions
group of six to eight people, he was at the very nerve
center of Dome's expansionist empire. He cut his teeth
on a few deals ranging from $100,000 to $1 million,
then quickly became involved in the "big time", which
meant marathon negotiating sessions during which all
possibility of sleep was forgotten and the participants
existed on endless cups of coffee and a great rush of
adrenalin.

One such occasion occurred during the very complicated takeover of Siebens Oil and Gas Ltd., which happened almost simultaneously with the TCPL deal in 1978.
Those working on the Siebens project went to work in
the Dome Tower in Calgary on a Friday morning, and
no one went to bed for two days. On the Sunday, McGrath
flew to Toronto with Bill Richards and a tax lawyer
from the Calgary legal firm of Bennett Jones (which
has a reputation on the oil patch for having the best
tax expertise in Canada, and dreamed up many of the
ingenious tax wrinkles for which Dome deals became
famous). Without rest, the Dome team headed into a
meeting of their own board of directors and then, joined
by others who had flown east on a later plane, began
further round-the-clock working sessions which continued relentlessly until late on the Monday. McGrath did
not get to bed from the time he awoke on Friday morning until he returned to Calgary on Tuesday. Buoyed
by the excitement of a complicated multi-million-dollar
deal, he found he could function remarkably well; fighting off tiredness, he was able to remain alert the entire
time—in fact he had difficulty sleeping when the chance
finally arose. This punishing regime was repeated several times during those hectic years as takeover after
takeover reached the critical stage.

Dome's financing of the Sieben's takeover was a *bra-
vura* performance, the financial equivalent of Luciano
Pavarotti tossing off nine high Cs in a Donizetti aria.
The deal came about when two managers from Canpar,
a subsidiary of the Canadian National Railways pension fund, one of the largest pools of investment capital
in Canada with assets in the billions, visited Calgary

seeking opportunities in the oil industry. Could anyone use a few hundred million? They called on several oil companies, but they had no need to go further than Dome, where they were received with special warmth. Siebens, a well-known takeover candidate with rich land holdings, had been on Dome's hit list for some time. The founding Siebens family held 46 per cent and the Hudson's Bay Company had 35 per cent; but with the Canpar connection Dome could work a little takeover magic. Canpar, not Dome, bought 100 per cent control of Siebens and then turned around and sold Dome 76 per cent of the *assets*. This could be done without Canpar paying tax on the sale since it enjoyed a special tax-exempt status. And Dome, since it was buying *assets* and not shares, was able to claim a fast tax write-off. Dome paid $400 million for its interest, comprising $396 million in notes and just $4 million in cash; it also undertook to manage Canpar's remaining interest in Siebens. The cleverness of this coup brought gasps from big-business fans from Vancouver to St. John's.

The Siebens acquisition gave Dome a dominant land position in Western Canada, adding 14.6 million net working interest acres and 23.6 million royalty interest acres to its already substantial land holdings. A large part of the Siebens acreage consisted of mineral titles held in perpetuity, originally granted to the Hudson's Bay Company in 1670. These lands were especially attractive because oil produced from them does not attract a provincial royalty.

Dome subsequently acquired a 65 per cent interest in the Canadian properties of Mesa Petroleum, an American company. This deal, like the Siebens one, was extremely complicated and used money from twenty pension funds made available through a company called Starvest, started by the two former Canpar fund managers. After this exercise the federal government called a halt to speculative oil industry investments by pension funds. Senior officials in the Ministry of Finance in Ottawa were furious with both Dome and the pension funds for exploiting the tax system so unashamedly.

Not everyone employed by Dome worked as compulsively as McGrath did in the acquisitions department,

but the stimulating environment attracted an unusually large number of workaholics. Getting to work for
Dome in its glory days was a bit like taking holy orders,
but the worldly rewards were better. To be a Dome
employee was to be someone rather special, a bit like
belonging to an elite regiment. By the time an employee
was hired, the company knew a great deal about him
or her. Anyone engaged for a professional, technical, or
supervisory job had to undergo rigorous psychological
testing, at least two interviews, a diligent reference
check, and, although they were not told about it, a credit
check. Many potential employees found the idea of psychological testing repulsive (particularly those from East
Europe), but most went along with the procedure, since
Dome was an attractive, vibrant company. The psychological assessment (known irreverently as the "shrink
tests") took, in total, about fourteen hours and was conducted for Dome by outside industrial psychologists.
The candidate would take home a package of tests requiring six to eight hours' work, and this would be
followed by a more intensive supervised series of tests
completed against the clock, which usually took four to
six hours. The object was to determine the rapidity of
a potential employee's mental process, and his or her
intellectual skills, such as the ability to learn, verbal
and numerical reasoning, practical judgement, and an-
alytical reasoning. They were also useful in detecting
sales skills (mainly the ability to sell ideas), supervisory ability, and such qualities as motivation, energy
level, cooperativeness, tact, and diplomacy. Usually only
the final two or three candidates for a job were asked
to undergo this scrutiny, which produced a seven-page
report that was regarded as only one tool in the company's hiring process. The assessments turned out to
be remarkably accurate in predicting an employee's
performance; as a cross-check on the effectiveness of
the shrink tests' predictive ability, more than one thousand test results were correlated by computer with sub-
jective managerial assessments of the persons' actual
performance after they had been in the job awhile. The
results were statistically very respectable. When Dome
was urgently seeking ways of trimming its costs in 1982,
an internal review suggested that the shrink tests be

scrapped to save money. They were retained, however, since the company's senior management felt they were well served by the psychological approach to hiring.

But not everyone could take Dome as seriously as it took itself. A senior manager of another company, razor-sharp and well connected politically, was approached by Dome, who wanted him to handle some delicate government relations work. The recruiting process was as follows: he would spend several days in the Dome Tower getting to know Jack and Bill and other senior managers; then there would be the shrink tests, to be followed by a formal employment interview. To his repeated inquiries as to what the job might be worth, he was told Dome would pay whatever was needed to get the right person. But did that mean $40,000 or $140,000? Again, a vague answer: whatever seemed necessary, but since the advice and intelligence the potential employee would be expected to give the company would influence decisions amounting to hundreds of millions of dollars, they wanted to be absolutely sure they had the right man. The manager, exasperated and unimpressed, declared that the whole prospect was far too much for his ego, and that Dome as an organization was simply "too flaky" for him to consider joining.

Traditionally the hallmark of a Dome person was his (or her) aggressiveness. The company's employees tend to be better educated than average and a little smarter, and they don't mind showing it. Over the years Dome has hired its fair share of rebels, people who don't fit into the restrictive mold of the multinational oil companies. Part of the fun of working for the company was that everybody was listened to, and if a person volunteered to take on extra responsibilities, chances were the work would be heaped upon him. By the late 1970s Dome had developed a remarkable—and revealing—hiring philosophy: there was simply no one it could not afford. Still, this did not mean that it went out "buying" people with outrageously attractive salaries, because it was felt that a person whose loyalty could be purchased could just as easily be tempted away. Although it would deny it to its competitors, Dome did, however, build a fine team of "head-hunters" who pursued talent wherever it was to be found around the world.

But the most aggressive executive at Dome, a man of awe-inspiring energy and boundless ambition, wasn't recruited by head-hunters. He rose through the ranks and rapidly reached the second-highest position in the company, at Gallagher's right hand. His name was Bill Richards.

Chapter Seven

BILL RICHARDS: BOARDROOM GUNSLINGER

The great driving force behind Dome's takeover of the Hudson's Bay Oil and Gas Company, and earlier acquisitions that saw the company change its style and grow explosively in the late 1970s and early 1980s, was Bill Richards, a man of restless energy and brilliant imagination. Ideas shoot from Richards' brain like fireworks, dazzling and beguiling those around him; occasionally onlookers are singed by the pyrotechnics.

William Edmund Richards was brought up by a widowed mother in Winnipeg's North End, and did odd summer jobs for the CPR. He served in the army, and then took a law degree at the University of Manitoba on an abbreviated course for Second World War veterans. Fellow law students remember him as being exceptionally studious as he tried to make up for lost time, but otherwise unimpressive. He had a rural legal practice in Cartwright, Manitoba, for a short while, but was never comfortable with the unwelcome intimacies revealed to a small-town lawyer: he disliked knowing everything about everyone in town because he filed their divorces, calculated their income tax returns, and settled their estates. Three years' practice was enough experience to qualify for admission to the bar in Alberta without articling in that province, and Richards took his chance.

"Things have been pretty damn slow for generations in Winnipeg. It's a nice place to live but it is sure hell to earn a living there. The degree of economic and social mobility in Manitoba is not very high, so my objective was to move to a more prosperous place, and Alberta was rather conspicuous in that respect," he says.

He joined Dome in 1957, for no other reason than that there was work available in the legal department, and he needed the security of a job straight away to support his wife and his two small children. "I always enjoyed practicing law, and my long-term objective was not to be in business but to go back to the practice of law," he says.

He quickly made his mark, however, by the cleverness with which he represented the company at various government hearings, where Dome sought approval to build up a profitable empire from the processing and transportation of natural gas liquids. His appearances before regulatory bodies have always been virtuoso performances. Many years later, a former member of the old B.C. Energy Commission remembers that the commission was glad to rearrange their schedule of public hearings on several occasions to accommodate a flying visit by Richards, so invariably forthright and informative was his testimony. As a witness he is entertaining to watch as he dodges and weaves his way around hostile questions with skill and bluff good humor. Even in those early days at Dome, Richards' ability was evident, and very soon after he joined the company, his talents were spotted by Jack Gallagher, who asked the younger man if he wanted to remain a company lawyer or if he would like to move into the business side of Dome's operations. He opted for business, for which he showed a rare affinity, and he bounded up the corporate ladder: to vice-president, executive vice-president, and then president in 1974.

Richards' dedication to Dome is complete and wholehearted. One executive says, "This company is Bill's life. I am sure you could come in the office on almost every Saturday or Sunday and find him in here at some time." A number of executives with young families have complained to him about his habit of calling meetings for 8 a.m. on a Sunday.

Richards is a fighter, a boardroom pugilist of awesome credentials. His bulldog appearance makes it easy to imagine him as a merchant adventurer at the court of Elizabeth I, and in fact he admires the dashing Tudors for their industry and intellectual strength, and has even decorated his ranch just outside Calgary in ersatz Tudor.

Although he can afford the best tailors, Richards cares nothing about his appearance. A joke circulated within Dome that Gallagher started giving away clothing vouchers to staff at Christmas to get his president into decent attire. Although he now wears the appropriate businessman's uniform of expensive dark suit and crisp white shirt, his short, stocky figure resists any hint of elegance or even mere tidiness: his tie is most frequently askew, his pants need pressing and his shoes repairing, and a shirt button may have popped, unable to contain the pressure of a thrusting paunch. He has a shock of untamed hair, and an expressive face that switches quickly from intimidating frown to engaging grin.

Coming to the city directly from his ranch, he has been known to hold meetings in his elegant office with fresh manure on his shoes soiling the broadloom. Insiders tell the story of how he once met a group of pinstriped Toronto brokers in the Dome Tower on a Saturday morning. They had flown about 1,500 miles to meet him, and were understandably startled when Richards appeared in cowboy dress and firmly planted his boots on the coffee table, telling his visitors, "You can have twenty minutes of my time; I have to get back to branding steers." Asked if this is a true story, Richards says he can't recall but admits it's in character.

He can be brusque or solicitous, as occasion demands. Once, wooing a party of Japanese investors, he invited them to his ranch for a barbecue and moved them to tears by hoisting a Japanese flag, not the easiest item to find in Calgary, atop a seventeen-meter pole. Surprisingly, Richards' dashing style—so greatly at odds with their own behavior—greatly impresses conservative Japanese businessmen who are accustomed to seeing the presidents of large corporations act with the stately dignity of a figurehead. An executive who has travelled with Richards in Japan says he typifies for the Japanese their expectations of a successful executive in the bustling Western business world. They are not a little impressed by his willingness to give quick, honest answers to their questions.

In Calgary, Richards operates like an expensive dentist, briskly running two or three meetings simultaneously, leaving a secretary or assistant to entertain or pacify visitors as he makes his rounds. This style of operation occasionally takes on the character of a French farce as Richards meets with different groups who must not know the identity of other actors with whom he is dealing.

If Canada could harness Richards' nervous energy the country would have no worries about self-sufficiency. Dropping into a subordinate's office for a two-minute conversation, he paces up and down constantly. At times he seems almost like a Hollywood stereotype of a hard-driving business executive, bounding from crisis to crisis, an ever-present telephone (sometimes even two) at his head. He admits to being a "telephone junkie", and was once interviewed on the Calgary CBC radio program *Eyeopener* by radio telephone from his car as he negotiated his way through traffic. He often arrives at the Dome hangar at Calgary airport in his Mazda sports car engrossed in a telephone conversation. With an absentmindedness more often associated with ivory-towered academics, he leaves the engine running and the door open after bounding from his car, and then grabs a telephone in the terminal office to make more calls while his executive jet is fired up; once aboard the plane, the telephone routine continues almost as soon as he has buckled his seat belt. (A business jet, his to command at an hour's notice, is just an extension of his office: the company's twelve-passenger Gulfstream II, which he uses on longer journeys, is equipped with an electric typewriter and a photocopier.) The Dome flight staff, appalled by the appearance of his car, routinely wash it for him unasked while he is away on trips, and since he is forever losing his ignition keys, they quietly had a dozen copies made. When he's not using the Mazda, he often drives to work in a pickup truck.

Almost everyone who has met Richards uses the same adjective to describe him: brilliant. His rapid-fire decisions are legendary. One executive, recruited by Dome from a major U.S. oil company, was asked to prepare an analysis of a business proposal. Although he insisted it would take him about a week to get the numbers together, the new man was called into Richards' office the next day and asked, "Well, what have you got?" With some reluctance, he offered his initial findings and was nonplussed when the president, after a few penetrating questions, made a decision involving a considerable expenditure in a few minutes. Said another executive, "The fellow came out just shaking his head. He said his previous company would have taken six months to make that decision, but it turned out to be a very astute one."

One Dome story about his deal-making style is almost certainly apocryphal. In February 1980, Dome agreed to pay $700 million for the oil and gas interests of Kaiser Resources Ltd. of Vancouver. Richards, impatient with temporarily stalled negotiations, was reported to have said, "Hell, let's give them another $100 million and get the damn thing over with."

He is motivated by an intense sense of competition. "It seems that Bill will do anything if it is competitive," says one Dome executive. "To him, the business world is a big game. It is like going into a prizefight; you don't have to do any favors to anyone until they are down. But Bill loves people and he will never beat up on a little guy." This competitiveness shows in his approach to personal fitness. Several times he has successfully lost weight—as much as thirty pounds—by wagering sums of money with other executives to see who could lose the most in a given time. Then the boring business of weight loss becomes interesting: it is a competition!

A number of Dome insiders have worried that Richards has too much on his plate and makes too many decisions personally. Says a former Dome executive, "One of his failings is that he doesn't delegate very well. He has too many people reporting to him. I always figured that eventually Bill would get his tit in the wringer and he wouldn't be able to extricate himself, for the simple reason that he tries to do too much for one man and he is bound to make a bum decision one of these days." Mixed metaphors aside, the executive's prediction turned out to be prophetic. Richards, though, doesn't accept the criticism that he tends not to delegate work. In his office there is a photograph of half a dozen people moving a heavy piece of equipment at his ranch while the denim-clad Richards looks on. He has captioned it: "Delegation." He claims he has so many people seeking his attention that "it makes it impossible for me to get into the nitty-gritty of their jobs. So as a result they do tend to have a lot more delegated responsibility."

Dome executives soon learned that while Richards is very approachable, they had better not waste his time. He is renowned for kicking people out of his office if they attempt to discuss problems without offering up their own answers. "I often tell people, 'Don't come to me with your damn working papers.' In other words, in an ideally run company, which of course doesn't exist, a person in my position should just be sitting back there and people would be presenting perfectly thought-out solutions and opportunities and I would just nod my head. Unfortunately, they don't. Sometimes they come forward with crap where they haven't thought it through, where there are obvious flaws in their reasoning. They haven't done adequate research. I am very impatient with that, simply because I think it is demeaning to them to come forward to management and say 'Well, here's a situation, what will I do, boss?' as if management has a magnificent superior knowledge to everybody else in the world. I feel they should come forward with a well-thought-out proposition and they should fight like hell to sell it as long as they are being rational, to the point that I have to shoot holes in it. I don't want to see tedious rows of figures. I assume the arithmetic is right and the work has been done properly. I want a proposal and an analysis and options and a reasonable review of the assumptions. My view of a well-run company is that you always delegate the maximum amount of responsibility to the lowest possible level."

Richards contends that he is not a hard taskmaster, but Harry Eisenhower, Dome's corporate secretary and a fellow lawyer, disputes this. "I don't know that Bill really takes cognizance of the time and effort that you may have to put in to do something for him. If he does, then he makes out that he doesn't. I think when he wants something done and wants some information, the people crack to get it done for him, and I don't think he appreciates sometimes how much time someone would have put in. But after working for him for many years, I greatly admire him."

Unlike Gallagher, who thrives on consensus, Richards doesn't encourage debate, and belligerently shoots down opposition unless the executive has the determination, and courage, to argue his case vigorously. Not all do: Peter Breyfogle, who went to Dome as vice-president of finance from Massey-Ferguson, whose European operations he had once run, was never able to win arguments with Richards, and eventually he left. By contrast, John Beddome, a heavyweight group vice-president in charge of Dome's exploration and production, whose ascending star is expected to place him in the president's chair one day, is made of tougher stuff. He is not afraid to tell Richards: "We'll discuss this when you are talking sense."

Ian Waddell, the NDP's former energy critic, suffered the humiliating experience of tangling publicly with Richards at an oil industry conference in Victoria, B.C., in September 1980, shortly after taking over the shadow portfolio. He labelled Dome with the NDP's familiar "corporate bum" tag.

"I was pretty naive," Waddell admitted later. "Richards made mincemeat out of me and the whole audience booed. I realized I had to get my act together." Much later, in Ottawa, Waddell enjoyed jousting with Richards. He compared his opponent's performance as he skated around tricky questions before a Commons committee to that of hockey player Wayne Gretzky, and commented wryly that Richards knew how to drill not only for oil in the Beaufort Sea but also for favors in Ottawa.

Despite the emerging portrait of a tough, even ruthless, hard-driver, Richards inspires affection among those who know him well, and they say that beneath a brusque and often domineering exterior there is a center that is decidedly soft. He insists on approving all company firings personally, and once spent three hours solicitously discussing the future of a junior employee he did not know but whose work had deteriorated because of serious personal problems. He agonized whether the company had done enough to help. This deeply felt concern for the well-being of employees came across strongly to Don Johnston, the company doctor, recruited in early 1981 after a career in the Canadian forces. At his initial briefing Richards told him: "We hire an awful lot of good people in this company, we try to hire the best, and we beat up on them, I suppose. That's okay as long as they can look after themselves, but we don't know when things aren't going well sometimes and that's where we want you to help. We certainly want to help if we are overloading them and things are going wrong and we don't know about it. Your terms of reference, in one line, are to look after the health of Dome." How Dr. Johnston tackled that was his concern. He got no interference.

Richards is contemptuous of grand designs and broad strategic plans in business and is openly and unashamedly opportunistic. "I have always believed in a fairly energetic approach to business. Growth in itself in my view is immaterial; what you should really be trying to do is searching for maximum profit and maximum value for your shareholders. One may say, 'Well, that isn't a strategy that has worked out very well for you recently,' and I concede that it hasn't, obviously for a variety of reasons—but none the less it has been our objective."

People frequently remark on the difference between Gallagher and Richards, superficially so obvious: while Gallagher is elegant and cautious in speech, a diplomat and a master of compromise, Richards is rumpled, outspoken, and pugnacious. His willingness to comment to the press on sensitive issues has caused him trouble more than once, paining the cautious Gallagher, who has tried repeatedly, but vainly, to muzzle him. The differences run far deeper, however. Gallagher is fascinated by science and technology and sees himself as an "explorationist" (to borrow an oil industry accolade), a dedicated searcher for the earth's riches; he is a visionary and cherishes the role of elder statesman. Richards, by contrast, has no interest in technology and has only once visited the Arctic, where his own company is

engaged in pioneering work that has been the toast of
the oil industry, arguing that it is far more sensible
and less time-consuming for him to be briefed by experts in his office than to traipse around looking at a
lot of machinery he doesn't understand. Field trips bore
him and so he avoids them. Richards is at heart a corporate acquisitor, and while he denies he is obsessed
by the imperative of growth, others within the company
say this is precisely his motivation. It is only since he
became president of the company in 1974 that Dome
entered its aggressive period of expansion by takeover.

There have been frequent disagreements between
Richards and Gallagher, with occasional rumors that
Richards has been fired or is about to quit. Such tales
are often exaggerated, although Richards definitely
threatened to resign early in 1982. At that time Gallagher was negotiating to bring in Robert Bandeen,
former president of Canadian National Railways, as
vice-chairman, a move that would certainly have
preempted any claims Richards may have entertained
for Gallagher's job as chief executive officer. The court-
ship of Bandeen, who had the type of solid credentials
required to placate the banking community, came to
nothing, however, and the stormy relationship continued. The two have, in fact, worked together remarkably
well over the years, despite their obvious differences in
temperament and approach to business. Gallagher says,
"If there is any friction, it's that I like to reflect on
major decisions: they don't have to be made overnight.
Bill's nature is to make a decision and move on to the
next thing. But I admire his many talents because he
is a very capable person, and I brought him on and
promoted him. Where we differ is in the philosophy of
how to build a company. I enjoy growing from within,
and up until a few years ago we did everything with
what we had generated ourselves. The acquisition approach is Bill Richards' initiative." Richards has said
of their relationship that "its having lasted almost a
quarter of a century is a great tribute to Jack Gallagher's patience."

Richards says he never worried about whether he
was Gallagher's heir apparent as the chief executive.
"I have enjoyed the job I have had. I have had all the

scope to get a good deal of satisfaction out of what I
was doing. I have never been very title-conscious. I am
an uncomfortable 'big shot', to be honest with you; I
don't enjoy the attributes of office too much. As a matter
of fact, they frighten me a little bit, because I have seen
men who have had these very elevated positions and
allowed themselves to take them very seriously, and
then when they are out of those positions they are
crushed because they depended too much on being
propped up by being a big shot."

Richards, who is fifty-seven, is a widower with four
grown sons to whom he is very close. One son runs his
ranch, which Richards manages to visit for a weekend
once or twice a month. "I can be as involved or as un-
involved as I choose," he says. He is involved with another of his sons in a real estate development in Calgary,
and they meet for breakfast or lunch several times a
week to make business decisions. "I don't really need
to do it, but it's my fun and recreation. I do it mostly
for the enjoyment, not the money."

He certainly does not need the money: he earns over
$268,000 in salary and benefits and owns 1.4 million
of Dome's shares, which at Dome's dazzling stock market high in the spring of 1981 were worth about $35
million, although by the fall of 1982 they had plunged
in value to 4.1 million. In common with all serious
businessmen, he also enjoys numerous perks, some of
which Finance Minister Allan MacEachen's budget of
November 12, 1981, moved to curtail. In reaction to the
news, Richards moved like a horse leaving the traps.
He was then enjoying an interest-free loan of $135,000
from one of Dome's subsidiaries to help him buy his
home. Under the new budget rules, such a loan would
have attracted tax on imputed interest of $23,176. Al-
though the contents of the budget were not broadcast
in Calgary until the early evening, Richards successfully hustled to repay the entire $135,000 to the com-
pany by midnight.

"If you look at my life-style," he says, "you'd probably
think of me as being a very possession-minded person.
The fact is I don't give a damn for possessions, although
I find that it's fun to have [business] projects to do."

His intense dedication to work leaves him little spare

time, but if he had more, he would hardly know how to enjoy it. He doesn't have much inclination for hobbies. He is a book collector, although he calls himself just a "casual browser"; his taste in reading runs to history, with a special interest in generalship. He also has a well-stocked wine cellar, and methodically makes notes about each vintage, although he says he isn't a true connoisseur. His great, abiding, all-consuming love is business, and with his nimble mind, outspokenness, and colorful turns of phrase, he is an entertaining exception to the rule that single-minded workaholic businessmen are a bore. He is quick-witted, impulsive, hard-driving, roughly spoken, and, at times, very funny.

But there have been many times when a chronic flippancy, which he has difficulty restraining, has irritated his colleagues. The sense of humor which seemed so appealing when Dome was soaring loftily was not much appreciated as the company's fortunes floundered. Irrepressible even at the height of the company's financial difficulties, Richards indulged in some black humor, telling a joke about a family in which the father was in jail and the mother and sister worked in a brothel. The mother was eligible for promotion to become madam of the brothel, but she was desperately worried about her chances if her dark secret should leak out: she had a son who worked for Dome! Richards also told—against himself—a joke that had become popular on the oil patch. Richards, so the tale goes, says to Jack Gallagher, "I've got some good news and some bad news. First the good news: We can buy Imperial Oil for $6 billion." Gallagher's eyes brighten. "Well, that sounds quite cheap," he says. "The bad news," Richards warns him, "is that they want $50 down."

When Dome acquired the oil and gas interests of Kaiser Resources in February 1980, shocked Kaiser employees woke to find they were under new ownership and gathered in some bewilderment and apprehension for an address by Richards. Says a Dome executive, "They felt that Edgar Kaiser [their former boss] had stabbed them in the back by selling out. When Richards comes in, someone asks, 'Can we keep flex time?'—an arrangement of flexible working hours they had enjoyed under their previous owner. Richards shoots back,

'You can come in any time you want on Saturdays and Sundays.' That did nothing to put their minds at rest."

His flippancy, a product of his unrestrained optimism, not callousness, was especially resented in the spring of 1982 when Dome was forced to lay off staff. Managers who had spent a gruelling day firing tearful employees were in no mood to hear lighthearted banter from their president, a fact he never seemed to appreciate. At one meeting when Richards was being particularly irritating, his executive assistant, Steve Savidant, grabbed his own tie and pulled it over his head theatrically in a hangman's gesture—an eloquent indication of what he would have liked to do to his boss at that moment.

In January 1983, as an especially important budget meeting approached, apprehensive executives passed along the word to their president that he had better cut out the humor—or he would face a riot. Richards behaved impeccably.

A more serious resentment grew within Dome among people who felt that Richards, with his impenetrable facade of optimism, had withheld the true nature of the company's difficulties from them for far too long. As a result, many of them had made very bad investment decisions. "It's not as if they wanted to benefit from insider trading, but a lot of people made personal financial plans because the bad news was kept from them. There is a great deal of bitterness over this. There was this little circle of people around Bill and virtually everybody else was shut out," says an executive.

Despite these shortcomings many people in the company were hoping that a reformed Bill Richards, whose talents are undoubted, would remain to head a revitalized company. By the spring of 1983, after many months of crises, tensions, and disappointments, nerves were rawly exposed in the Dome Tower, accumulated misunderstandings and recriminations took their toll, and relations between Richards and Jack Gallagher deteriorated alarmingly. Observing this, the Dome administrative ranks began to polarize into distinct Gallagher and Richards cheering sections.

When the finger was being pointed publicly at Richards as the cause of Dome's problems, he refused to become despondent. "Oh hell, I don't mind [the criticism]. People are all for risk-taking and that kind of stuff as long as nothing ever turns bad on them. For a variety of reasons the HBOG thing has not worked out very well. I would not try to avoid a very large degree of responsibility for the initiative. How you allocate that responsibility, who can say." He had not been asked to resign, he said, although there were times when he felt like leaving on his own initiative.

"In terms of what the future holds, I don't know. My view is that I am the president of the company and I am going to see this thing through. I have no intention of participating in the creation of a difficulty and then running for cover. So it is my intention to do everything within my power to solve the problems and I think there is an awful lot we can do. Maybe we will have a new board of directors and they won't like me so well or maybe I won't like them so well and that will change. Right now, day by day, I am there and I am acting as if I am going to be there forever."

3

DOME AND THE NORTH

Chapter Eight

OUT IN THE COLD

The presence in the Arctic of an industrial armada of great technical complexity under the Dome banner in the 1980s can be traced to the excitement of an undergraduate geologist paddling a canoe along the rivers and lakes of northern Manitoba and the Northwest Territories almost half a century earlier. Jack Gallagher's infatuation with the resource potential of the North, stirred during those summer expeditions as a young man, never left him, although the dream long lay dormant. Towards the end of his globe-trotting career with Standard Oil of New Jersey in the late 1940s, his interest was rekindled, and this time it acquired a more specific focus. On loan from Standard to Imperial Oil (its Canadian subsidiary) for four months on what he called "a roving commission", he went into the MacKenzie Delta, and returned south itching to explore for oil there. He had seen studies by the Geological Survey of Canada and also the results of some early Imperial Oil field work, which indicated the region's great hydrocarbon potential.

He tried, unsuccessfully, to get both Imperial and Standard to share his enthusiasm for a major exploratory push, claiming that anywhere else in the world the oil executives would be interested in the potential of the Delta because it had all the attributes that a geologist looks for in a potential oil and gas basin. His argument was that major oil fields have been found in many of the world's great river deltas—those of the Mississippi (the Gulf of Mexico), the Niger (in Nigeria), the Orinoco (in Venezuela), and the Tigris and the Euphrates (the Persian Gulf)—and that the MacKenzie

Delta and the adjoining offshore area of the Beaufort
Sea have the same Tertiary sediments in which sixty-
five per cent of the world's oil and gas is found.

Frustrated at the cool reception his ideas received,
Gallagher became convinced that the multinational oil
companies would not be the ones to explore and develop
Canada's Arctic oil potential; the work, he believed,
would have to be undertaken by more aggressive, independent companies. (In this he wasn't entirely correct, for both Gulf and Imperial are now active in the
Arctic; however, they move far more cautiously than
Dome.) Reflecting much later on his difficulty in stimulating interest among his former employers in the Arctic in the late 1940s, he said he did not truly understand
then the problems of running an international oil company. A professional manager has to maximize his profits and invest where he can see the best economic return.
And it is difficult for such men also to be "industrial
statesmen"—a role in which he clearly sees himself—
because they would not remain in their jobs for long.

Gallagher felt the only way he could fulfill his ambition to help open up the Arctic was to do it as an
independent, and the offer to head Dome Petroleum was
the chance he needed. At Dome he was still a professional manager, but he was working for an indulgent
board of directors. By normal investment criteria, drilling in the Arctic was unthinkable. But Dome was not
seeking a quick profit: the company was looking perhaps twenty to twenty-five years into the future before
it could begin to get its money back. Even if it discovered oil and gas, it would take years to establish sufficient reserves to justify the horrendously expensive
transportation systems necessary to carry that production to markets thousands of miles distant.

In 1958, after seven years spent building the fledgling company, Gallagher attended a geological conference and heard news that set his pulse racing. Charlie
Dunkley, a retired Dome vice-president, recalls Gallagher's excitement: "Jack came back all full of piss
and vinegar." The Deputy Minister of Northern Development had told him about some work that the Geological Survey had been doing up in the Arctic Islands,
where the government geologists had been amazed at

the unusually promising geological structures they found and the great thickness of sediments there. Typically, Gallagher went off to look for himself, flying in a small twin-engined plane that seemed to back up in the wind. He was very impressed, and subsequently spent ten days in Ottawa with one of the company's geologists, looking through a stereoscope at aerial photographs and picking out areas on which he wanted to file for drilling permits.

Eager to beat other companies to the draw, Gallagher flew next to New York, where Dome held its board meetings at that time, and made a pitch to his directors for permission to file claims on an enormous area of the High Arctic, about 3.5 million acres in total. With some difficulty, he extracted reluctant approval from the board, which worried that Dome, still a small operation, would look foolish staking out land practically on the roof of the world, with no other oil company literally within thousands of miles. But Gallagher argued persuasively that if Dome wanted to move, it ought to do so ahead of the crowd, to seize the most promising areas quickly before the major companies moved in.

He then took a ten-day vacation in the Bahamas, his first holiday in seven or eight years, but as he was heading back to Canada he was asked to call Dome's chairman, Cliff Michel, in New York. There was bad news. Michel, who had served in the U.S. Navy during the Second World War, had talked to some of his former military contacts and also to a major Norwegian shipping company that operated oil tankers. Their advice was so disheartening that Michel sought to dampen Gallagher's enthusiasm: it would be seventy-five years before Dome would ever get a ship in and out of the Arctic, and the move was at least a century ahead of its time. Gallagher argued vigorously, but, at his chairman's bidding, finally agreed to delay filing for land in the Arctic until another oil company showed confidence in the area and filed first. He bided his time impatiently, and the moment word of the Arctic's hydrocarbon potential leaked out and another company acquired a land position in 1959, Gallagher stepped in smartly and filed for Dome within forty-eight hours. To his intense frustration, some of the land he liked best had already been snapped up, and his unhappiness was intensified when that area was found to contain enormous reserves of natural gas. (However, owing to its remoteness from markets, not so much as a thimbleful of this gas has yet been sold.)

In September 1961, Dome became the first oil company to drill an exploratory well in the Canadian Arctic, when it sank *Dome et al Winter Harbour No. 1* on Melville Island. Public interest was aroused, and a picture of the well made the cover of *Life* magazine. It turned out to be a dry hole, but that was unimportant. Dome had proved conclusively that it was *possible* to drill for oil on Canada's unforgiving northern frontier, many thousands of miles away from normal support services, and the drilling also provided useful subsurface stratigraphic information that assisted in planning future geophysical surveys and exploratory drilling. Typically, the bill was paid mainly by investors seeking tax shelters (soon to be a recurrent theme in Arctic oil exploration). Thanks to farm-out arrangements it had made with other investors, Dome actually only contributed ten per cent of the cost but obtained very substantial work credits from Ottawa which were sufficient to hold the company's lands into the mid-1970s. (Companies are obliged to surrender exploration lands if they do not perform a minimum work obligation; since no one, Ottawa included, knew if it was even possible to drill in the Arctic, the work requirements were not onerous in the early days.)

By 1967, the Winter Harbour experience to its credit, Dome was the leader in High Arctic oil technology, and it was leaned upon to become the operator for a new consortium, Panarctic Oils, put together by geologist John Sproule. Gallagher didn't want to do the work, because Dome had only a modest interest in the Panarctic group, and involvement there would preclude his company from working on its own lands, but he capitulated under pressure from Inco, Noranda, and the federal government, also a Panarctic partner, and reluctantly agreed that Dome would run the operation for three years until the new company could recruit and train its own staff. It was during this period that the first major Arctic gas discoveries at Drake Point and King Christian Island were made by Panarctic.

In the next few years Dome participated in drilling eleven wildcat wells on its own lands in the Arctic Islands and made three gas discoveries. American gas transmission companies, looking to future sources of supply, paid most of the bills. Dome got $30 million from Columbia Gas Development Corporation of Delaware, and then a similar sum from Consolidated Gas Supply Corporation of Pittsburgh, Texas Gas Transmission Company, and Panhandle Eastern Pipe Line Company of Houston. The Americans earned an interest in Dome's lands, plus preferential claims on any gas discovered.

Dome's strategy was to acquire large land holdings early, conduct the initial geological and geophysical surveys to establish potential values, and then, by disposing of part of its interests, have others pay most of its exploratory drilling costs, which it could not afford to undertake alone. At the end of 1981, Dome had a gross working interest in 15.7 million acres in the Arctic Islands and a royalty interest in 17.4 million acres. It also retained a nine per cent interest in the Panarctic consortium and was a major participant in the Arctic Pilot Project, led by PetroCanada, which was planning to liquefy natural gas from the Drake Point and Hecla gas fields on Melville Island for delivery year-round by icebreaking tanker, probably to markets in Europe.

In 1967, while still working in the High Arctic, Dome filed on a substantial acreage in the Beaufort Sea in the western Arctic, an area of the world with which it was to become synonymous in the public mind. Drilling techniques in the High Arctic differ considerably from those that were developed subsequently in the Beaufort. Some of the early wells in the Arctic Islands were sunk from land, but later operations were carried out in winter from rigs positioned on the thick ice which surrounds the islands for much of the year. In the Beaufort, the offshore drilling is done from drill ships, and the picture of red-and-white vessels braving the elements in the frozen North in the search for oil and gas riches has become a powerful symbol of Dome's enterprise and daring, as distinctly Canadian as the maple leaf and the beaver.

Ironically, the huge fleet of more than forty ships of various types and sizes that Dome assembled in the Beaufort was created by default. Originally the company didn't want to get entangled in offshore drilling, which it didn't understand and in which it had no expertise. It wanted to direct operations, but planned that others more knowledgeable would do the actual work; and so, in the late 1960s, Jack Gallagher sought help from the colorful Hunt brothers, Bunker, Herbert, and Lamar, who ran the largest independent oil company in the United States, started by their father, a highly successful wildcatter. The Hunts, who later gained notoriety by proving it was impossible to corner the world silver market, had an excellent drilling contracting company which they were starting to move offshore.

Gallagher arranged that the Americans would earn an interest in Dome's Beaufort acreage by drilling the first well, to be started in 1973 and completed the following year. The Hunts, however, had no idea of the complexity or the cost of what they were becoming involved in as they began making plans to mount a land-based rig on an armored barge. When Dome and Hunt signed their deal, the federal government had not clarified its regulations with regard to drilling in the Beaufort. Soon, however, the Hunts learned that Ottawa required extensive environmental studies, and would insist on the use of two complete drilling systems (so they could drill a relief well quickly in case of an oil well blowout), as well as support facilities such as icebreakers and supply boats. By late 1972, faced with two or three times the cost they had originally contemplated, the Hunts decided to back off, leaving Dome in a dilemma, having lost five years of the twelve-year life on its drilling permits.

At this point, Dome decided that, as a Canadian company, it would be easier for it to deal with Ottawa on the sensitive issue of Arctic drilling than for the Hunts; Dome would conduct the required environmental studies and secure the drilling permits. Rather than sue the Hunts over the broken contract, which would have meant lengthy litigation, Gallagher negotiated a new deal with the brothers: Dome would drill the first Beaufort well at the Hunts' expense, and would earn the option to acquire an interest in substantial additional acreage that the Hunts held in the Beaufort.

Still seeking outside expertise, Dome next approached Global Marine of Los Angeles, which had originated the art of drilling from a floating drill ship and was an acknowledged world leader in its field. Dome proposed that a new drilling company would be set up as a three-way venture: Dome would own half, Global would have a quarter interest, and Gulf Canada, which also has lands in the Beaufort, would take a similar interest. Dome executives spent weeks in Los Angeles working on a detailed document of agreement, but Global played their hand too shrewdly. They wanted the new venture to pay off the cost of the drill ships over the life of a five-year contract and they wanted a profit on top. Gallagher felt that the only effort Global was prepared to make was to walk around the corner to the bank! He reasoned that Global, from whom he anxiously wanted to acquire offshore expertise, would not be willing to put their top people to work on a project where they were not exposed to any monetary risk and had only a twenty-five per cent participation, and so talks were broken off.

The company suffered another setback when Gulf Canada was unable to persuade its American parent that it should participate in the Beaufort venture with Dome, apparently because of the absence of clear federal oil and gas regulations and the lack of assurance that they would have continuity of drilling in the Arctic owing to serious environmental concerns. (Beaufort operators still have no assured continuity of operation, although each year the process to obtain government approval of their drilling plans is less arduous.) Gallagher sympathized with the logic of Gulf's position: his own decision was rooted more in emotion than in reason, he said. But he believed that, as a Canadian, he could talk Ottawa around.

In March 1974, Jean Chrétien, who was then Minister of Northern Development, wrote to Dome saying that if the company completed environmental studies to the government's satisfaction it would be given the drilling permission it sought. Dome was now in a Catch-22: the extensive environmental studies demanded by the government would take about two years, but the company would have to make an early start to prepare for drilling or risk losing its handsome land position. Gallagher took a gamble that was truly monumental: with no guarantee that Ottawa would let him drill in the Beaufort, Dome committed about $100 million to the purchase and modification of two drill ships. It also committed itself to the building in Vancouver of four icebreaker-supply boats, each worth $8 million or so. Gallagher was on his way, and determined not to let anything stop him.

In mid-1974, the phone rang in the European head-quarters of Mobil Oil in London for Gordon Harrison, a Canadian and the company's planning manager for Northwest Europe. It was the type of telephone call senior executives receive often: a request from an acquaintance charged with selecting candidates for an important job. But if this inquiry was routine, the job description was quite out of the ordinary. Dome was searching for a seasoned manager to head a new group that would be responsible for drilling operations the company was planning, still without government clearance, in the Beaufort Sea, 350 miles north of the Arctic Circle. Could Harrison suggest any names? He was busy and not very interested but said he would look around. A month or so later he was approached again, and he suggested a couple of people who might fill the bill. This time, however, his own interest was aroused, and he phoned across the Atlantic to Calgary to learn more. By coincidence, he was told, Jack Gallagher happened to be in London, so the two men had breakfast together. They hit it off well, and a month or so later Harrison visited Calgary, had a further chat, and made a deal to become president of Canadian Marine Drilling Ltd. (Canmar).* At that point he did not clearly understand

---

Most of Dome's Arctic operations are technically run under the Canmar banner, but since it is a wholly owned subsidiary, no effort has been made to differentiate between the two companies.

what was ahead of him, although obviously it was a big job.

Like Gallagher, Harrison was born in Manitoba, where his father owned a farm, but since times were tough in the Dirty Thirties, the family moved to Stewart in northern British Columbia when Gordon was six, and his father went to work in a gold mine there. Stewart is an isolated community, and so Harrison had to tackle most of his high school work by correspondence, which system he doesn't recommend. By the time he reached Grade 11 he was able to attend a high school on Saltspring Island, just off Vancouver, and later he went to the University of British Columbia, where he graduated in 1954 as an electrical engineer. He worked for several oil companies in a variety of engineering jobs before settling down for a while with Mobil. As the company's operations manager, he began drilling operations on Sable Island off the east coast in 1968, and then started Mobil's offshore drilling work in the Atlantic in 1971, before moving up the corporate ladder to New York and then London. He felt like a pawn on a checkerboard, however, since the management of an international oil company is shifted around the world continuously and unsettlingly, and so he was ready to move back to Canada when the Dome job arose.

He did not see much of his new home in Calgary at first; from the moment he joined Dome in late 1974 he practically lived on an airplane for six months. Dome had done some preliminary engineering work for its Beaufort operations before Harrison arrived, and a few people had been assigned to the project temporarily. The shipyard contracts had not been given and the general construction program had not been totally authorized and put in action. Harrison had four major problems to worry about: he was responsible for the building of the drill ships in Galveston, Texas: he had to oversee the construction of icebreaker-supply boats in Vancouver; he felt it was vitally important to understand the inner workings of Ottawa and the status of Dome's drilling approvals; and he had to recruit a staff. Harrison recalls that when he awoke each morning it was difficult to decide what the day's priority ought to be. Added to the burden of work was the nagging uncertainty of whether Ottawa would, in fact, give the company permission to drill in the Beaufort. As it turned out, Dome's ships were actually going around the north coast of Alaska before the cabinet finally acted.

Harried though he was, Harrison flew over to Scotland whenever he could spare the time, to cruise around the beer parlors of Aberdeen. This may seem bizarre social behavior for an overworked executive, but it was all part of the job. Since he needed to find the nucleus of an organization with offshore drilling expertise, he headed for the obvious place for a talent hunt. The North Sea was then at the peak of its activity and the main service center was Aberdeen, and in its busy pubs off-duty drilling crews could be tracked down and interrogated.

At first Harrison felt he did not have much credibility; it was hard to convince anyone that Dome—of which many people had not heard—was big enough to take on the Beaufort project, or that the prospect of drilling in the Arctic was actually feasible. In addition, Harrison was handicapped by a lack of contacts in the oil contracting business, for while he had spent many years in the oil industry he had always been a company man, and he was now dealing with a specialized area in which he had no first-hand experience. But he persisted and soon developed a technique: if the same person's name occurred in several conversations, the chances were that he was worth interviewing. Finally he had a break-through when he recruited the superintendent off one of the North Sea rigs, Oran Monteith, a natural leader, who brought along about twenty people—sub-sea engineers, drillers, mechanics, and electricians—with him to Dome.

Harrison knew that the oil industry in Calgary believed Dome was engaged in a reckless project, and more than one former colleague had looked at him as though he were mad. He says it was the conventional wisdom of the Canadian oil and gas industry in 1976, when Dome began placing its first drilling systems in the Beaufort, that not only was effective exploratory drilling doubtful there, but the technology for production and transportation was, at best, a twenty-first-century possibility.

An illustration of Harrison's attitude to obstacles can be gleaned from a speech he gave to a marine conference in 1981, in which he discussed the historic voyage of the U.S. supertanker *Manhattan*, which travelled the Northwest Passage in 1969 to test the feasibility of transporting oil from the North Slope of Alaska to markets in the south by ship.

It showed bold leadership and clear authority are essential to advancing man's capacities and enlarging mankind's horizons. The management of Humble Oil Company [which operated *Manhattan*] were forward-looking, confident, thrusting and marching men. On discovery of the giant Prudhoe Bay oilfield in Alaska in 1968, *they spent little time dithering with desk "feasibility studies" and cluttering their minds with doubt.* They simply accepted advice that commercial shipping was feasible through the Northwest Passage and set about to prove the advice was correct. In record time they converted a conventional tanker to an armoured icebreaker and conducted a bold and fascinating Arctic marine experiment. [Italics are author's.]

Dome too has never wasted time "dithering with feasibility studies", or cluttered the mind with doubts.

Gallagher's $100 million gamble is grudgingly admired as an example of entrepreneurial daring by Dome's harshest critics; it is not something that Imperial or Shell—and certainly not a Crown corporation like Petro-Canada—would dare take on. But Gallagher says he had complete faith in the people in Ottawa that, if Dome conducted the tests as it had promised, he would secure his drilling approvals. Still, his gamble nearly turned sour; when a new environment minister, Jean Marchand, was appointed, practically the first statement he made after taking office was that he would not permit Dome to drill in the Beaufort in 1976. Marchand was responding to intense pressure from aroused environmentalists who feared the devastating impact that an oil-well blowout could have on the particularly fragile Arctic ecosystems. Gallagher's bland account of what happened next is fascinating in what it leaves unsaid: "So I personally went back to Ottawa with Gordon Harrison and we met with eleven ministers individually in the space of two days, and we effectively reversed that decision with the help of everybody involved."

About ten days before the fleet was to sail for the Arctic, another problem developed and Gallagher had to use his considerable persuasive powers again, this time in the United States. Natives from the North Slope of Alaska were worried about the possible effects of an oil spill or well blowout on their fishing grounds, and this discontent flowed through the Alaskan state government to Washington, D.C., with the result that U.S. Ambassador Thomas Enders began making hostile noises in Ottawa. Gallagher and Harrison talked to Enders; then, alarmed at his unfriendly attitude, they got Gallagher's contacts in External Affairs to arrange a series of important appointments in Washington, D.C., where they met with various interest groups. They treated the representatives to one of Dome's slick slide-shows on the results of their environmental work, how they planned to drill, and the efforts they were taking to protect the environment. Finally, five agents from the U.S. government flew out to see the ships for themselves, and left apparently convinced that Dome was taking all the practical precautions possible. Gallagher, the super-salesman, had closed another deal.

Each year since the mid-1970s Dome added significantly to the Canmar operation, and by 1983 the value of its Beaufort fleet and associated land facilities approached $690 million. Dome may have got into Arctic drilling by default, but having dealt itself an unwanted hand it played its cards astutely, quickly making the very interesting discovery that an oil company doesn't actually have to *sell* oil to make money. It came upon a "gusher" of a different kind—drilling wells under contract for other companies. Although Dome has yet to produce a single barrel of crude commercially from the Arctic, it has generated substantial profits from its Beaufort Sea operations. The fleet, operated by its subsidiary, Canadian Marine Drilling Ltd., cornered the market for a number of years as the only outfit able to drill in relatively deep water in the Beaufort, and it took full advantage of the opportunity that this monopoly presented. In view of its reluctance earlier to dance to Global Marine's tune, Dome's own charges for third-party drilling are noteworthy: its fees are based on a thirty per cent return on its investment plus an operating charge, and the contracts are written so that it will get its money even if the drill ships are trapped in harbor all summer by unusually severe weather conditions. (As early as 1974, before the fleet was built, Dome was already hedging its bets, and it received $10.4 million from others holding land in the Beaufort as prepayment towards future drilling.)

Canmar's earnings began even in its first extremely brief 1976 season when it made $26.8 million; the following year they jumped to $63.2 million. Then, with the exception of a dip in 1978, they grew steadily until 1980, after which they began to soar. In 1981, in response to the "Canada first" incentives of the National Energy Program, all of Dome's exploration activities in the Arctic were taken over and paid for by its newly created subsidiary, Dome Canada, which Dome Petroleum managed. Acting as a contractor, Canmar, still 100 per cent owned by Dome, did all Dome Canada's drilling but billed it in exactly the same manner as it would with any other third party for whom it worked (such as Gulf or Hunt). As a consequence, Canmar's profits (and the cash that flowed back to Dome Petroleum) rose dramatically: it made $130.5 million in 1981, about $186.6 million in 1982, and was expected to earn $214 million in 1983. These earnings would not have been possible, though, if the company had not been willing to plow back its profits into a greatly expanded Arctic operation, with the resulting ability to perform additional work both for itself and for others. Dome took both the technological and the financial risks, while other companies who have used the drilling systems have only been willing to sign contracts for use of the equipment on a year-to-year basis.

While such a handsome cash flow is received eagerly by an impecunious parent, the operation of the Canmar fleet has a long-range impact on Dome's affairs that may prove to be even more beneficial than ready cash. The fleet's monopoly gave Dome a splendid horse-trading advantage. When it drilled on another company's land, it almost always levered a piece of the client's property as part of its fee, and thus built up its Beaufort Sea acreage, potentially the source of great riches. In eight years from 1974, Dome *doubled* its extensive Beaufort land holdings.

Although Dome has built a very substantial operation in the Beaufort at great cost and acquired an enviable land position, it is often overlooked in the fuss that surrounds the company's Beaufort activities that over the years only about ten per cent of Dome's annual exploration budget has been spent there, the bulk of its exploration work still taking place in the safer and far less expensive fields in the western provinces of Canada and in the United States. Certainly, enormous sums have been expended in the Beaufort, but Dome became expert at spending other people's money—that of other oil companies, private investors, and also the Canadian taxpayer, through generous tax concessions and later through outright grants.

Dome has been remarkably successful in recruiting talented risk-takers, people who are willing to experiment and to be technically innovative, and who regard difficulties, however daunting, as interesting problems to be solved and brushed aside. This sounds dangerously like a company-written testimonial; it is, however, a widely held oil industry view of Dome. There is another side of the coin, however, and it is less flattering: Dome's risk-takers and innovators were also excessively impatient, impulsive, and spendthrift, worrying far less about the control of costs than about potential results, as they forged ahead in their compulsive desire to extract hydrocarbons from the Arctic.

In part, this free-spending attitude was forced upon Dome by the very difficult logistics of its operations, with activities located almost 1,300 miles north of Edmonton, the nearest place where standard oil-field supply equipment services are available. Drilling seasons were ridiculously brief, and so the company could not afford to lose a single minute in idleness, let alone the hours or days it would take for a needed piece of equipment to be flown in from Edmonton. Large inventories (with frequent overlapping in the ordering of supplies) were kept at readiness in the Arctic where they would be immediately available. If the need was urgent enough, a manager would call for a helicopter or a company jet with about as much thought as his counterpart in southern Canada would demand delivery by pickup truck. Ship's captains and chief engineers had authority to spend up to $10,000 without higher approval, although this privilege was removed when Dome began to tighten its cost controls in 1982. By then the chaotic supply system of the early years had been replaced by more orderly computer-based procedures. Still, a ship's chief engineer, contrasting the attitude of Dome with that of a previous employer, recalled that he had once asked the latter company for ten feet of piping and had been told he could only have six. "Hell, here at Dome I could ask for a thousand feet and get it, no questions asked," he said.

Bold decision-making and the rapid execution of ideas have characterized Dome's operations in the Beaufort. Everyone involved attests to the stimulating atmosphere, where even the most eccentric ideas were scrutinized earnestly for possible merit—such as the idea (discarded) that Dome should buy a Second World War aircraft carrier from the U.S. Navy for use as an operational base. ("Just think, we would have lots of storage space and we could land Twin Otters on the deck. Hell, we could have kept the guns, too," an executive joked.) Dome has rushed headlong and with such obvious zest into its Arctic operations that its competitors argue many of its engineering operations are undertaken with little more than superficial planning. One rival executive remarks: "Dome have a terrible habit of constructing something before they engineer it," in other words before they design it in exacting detail. "From my perception you engineer the hell out of something and you have a good handle on costs and surprises. You try to minimize surprises, because most of the time surprises are unpleasant. Dome have a tendency, because they are a very aggressive, high-risk, impatient group of people, to put their money where their mouth is before they do the engineering. They can look at us [a conservative multinational] and say, 'Those bastards, all they do is engineer the thing to death before they make a decision.' And it's true you can engineer things until hell freezes over and never make a decision. [But] Dome do a lot of things on a back-of-the-envelope calculation. We would only make the same decisions after a very exhaustive analysis, and maybe have a third party criticize the analysis. Ours is a very risk-free decision-making process, but it is also slower, so there is a trade-off."

Dome sought advice about the problems of drilling in the Beaufort from Imperial, which had earlier experience both in the Mackenzie Delta and closer inshore in the Beaufort Sea, but in their great enthusiasm to press ahead urgently they ignored many of the warnings they were given, and as a consequence encountered numerous technical difficulties, including the risks of over-pressurization of wells which they had been warned about. They learned quickly, however, and corrected their early mistakes in short order. Dome's style, a product of unbounded enthusiasm, is to solve problems as they occur rather than engage in tedious time-consuming planning. Problem-solving is great fun to an engineer, and Dome thrives on adversity. Just one example illustrates the difference between Dome's "damn-the-torpedoes" approach and the ultra-conservative progress of a giant like Esso Resources (Imperial's exploration arm). Dome built its Tarsiut artificial island in just ten months against enormous odds and in the face of fierce early-winter storms. The island suffered severe damage and Dome spent many millions making repairs. When Esso used a similar island-building technique, it had a $55 million steel caisson constructed in Japan and towed to Tuktoyaktuk, where its huge investment was allowed to sit idle until the ice cleared from its drilling location in the summer of 1983, instead of attempting to place it in the fall of 1982 and run the same risks that Dome encountered.

Dome's pioneering role in the Beaufort Sea is no longer scoffed at by the oil industry, and many others are now searching for petroleum riches in its wake. Dan Motyka, vice-president of Gulf Canada Resources Inc., says bluntly: "I think it is because of people like Jack Gallagher and Gordon Harrison that we in the Canadian petroleum industry are actively pursuing the Beaufort area. If the world had waited for very conservative companies like Gulf, the world would still be waiting. Thanks to the courage or stupidity of Dome— call it what you will—we are in the Beaufort too. Those guys had the balls to get us into the Beaufort and they deserve a considerable amount of credit for it."

Chapter Nine

BEAUFORT:
FACT OR FANTASY?

It quickly became apparent that there is oil under the Beaufort Sea; in fact, its presence was never much in doubt. The key question, still unresolved in 1983, with Dome in its eighth drilling season, is *how much* oil will be discovered and whether there will be sufficient to pay for the very expensive transportation systems needed to carry it to market. As early as 1977, at the start of its second Beaufort season, Dome was making flamboyant claims of a potential oil field to rival the big Alaskan discovery at Prudhoe Bay, which would mean enough oil to supply Canada's entire energy needs for several decades. This was big talk indeed on slim evidence, and caused many eyebrows to shoot up. Simply by estimating the volume of sedimentary rock— potential hydrocarbon traps—lying in the Beaufort, and comparing them with other prolific oil-producing areas (such as the Gulf of Mexico), Dome made extremely optimistic assumptions about the quantities of oil that might lie under the Beaufort. On such slender threads did Dome suspend the weighty illusion of riches before the Canadian public. Six years later, Dome remains the most publicly optimistic by far of the oil companies engaged in Beaufort exploration. It is convinced that it will find an "elephant"—a truly enormous oil reservoir—or perhaps even a whole herd of "elephants" under the icy waters of the Beaufort. It is noteworthy that this view is held privately by several "explorationists" at Gulf, although Esso tends to discount the theory, believing rather that the Beaufort may contain numerous smaller pools of oil of marginal commercial value.

The great frustration of exploring the Beaufort with drill ships is the extreme shortness of the season: the vessels cannot leave their winter harbor at McKinley Bay on the Tuktoyaktuk Peninsula until about mid-June when the ice begins to clear, and they must return about four months later, so it is virtually impossible to drill and complete the testing of a well in any one season. This is also the reason Dome shoots from the hip when it makes operational decisions; with the briefest of summer drilling seasons, ponderous analysis and reflection are unaffordable luxuries. In its first six years of Beaufort operations, Dome was limited to less than five hundred days of deep drilling. Once it began building islands and using semi-submersible drilling caissons, which, unlike drill ships, can survive the onslaught of winter ice, it began drilling year round, a vastly more efficient practice. A consequence of its early pattern of drilling activity, limited to less than half the year, was a trickle of tantalizing information released at the conclusion of each season. Several oil discoveries promising great riches were announced, although definitive testing was incomplete, and the audience had to tune in repeatedly to see how the plot, like a long-running adventure thriller on the radio, was developing. Dome's announcements often seemed to hint at greater discoveries than those revealed by its more cautious competitors, prompting some to worry that Dome might be pumping up the value of its stock by overstating its results, and there were indeed times when Dome was desperate for good news from the Beaufort to help it survive its acute financial difficulties.

The analysis of preliminary oil discoveries is an area where only a handful of people, expertly qualified and with access to confidential data, are competent to make a proper judgement. Part of the difficulty is that the interpretation of oil-field reserve data is a marriage of art and science, and no two reservoir engineers are likely to produce the same results, even if they are working with identical figures. Dome says it has a knowledge of Beaufort geology superior to that of other companies (a disputed claim), and that it has developed better interpretational techniques. Dome is also more optimistic than others in its view of the percentage of oil, known as the recovery factor, that it expects to produce from a given well. The earth does not readily relinquish its hold on the oil contained beneath its surface; generally only about thirty per cent of an oil pool can be recovered using normal production methods, although this fraction can be enhanced by secondary recovery techniques, such as water or gas injection, or exotic tertiary treatments such as fire flooding—the creation of controlled underground burning—to force the reluctant oil to the surface. Such matters of interpretation are of more than academic interest; if a company persuades its backers it can recover forty per cent of a field rather than twenty per cent, this adds enormously to its assets and hence to its ability to borrow against future production. The implications for a financially troubled company like Dome are obvious, and to resist the overwhelming temptation to overstate its case would have required a saintly rectitude.

Because of widespread skepticism about its claims, Dome opened its books in the fall of 1981 to an American company of reservoir engineers, Degolyer and MacNaughton of Dallas, known throughout the oil business for their toughness and high-minded integrity. It was like asking for a character reference from the Pope. The American engineers examined technical information relating to two areas in the Beaufort where Dome had found oil, and concluded that the Kopanoar structure had potential oil accumulations of between 1.8 and 4.5 billion barrels and that the Koakoak structure had the potential of from two to five billion barrels. The Americans' opinion of the recovery factor was agonizingly vague, ranging, they said, from fifteen per cent to forty per cent under existing engineering and operating techniques. They added that there were several other structures, more or less similar, in the same general area of the Beaufort where Dome has majority interests, most of which had not been drilled. The report settled nothing, for at the low range of Degolyer and MacNaughton's projections there was so little oil it would have to remain under the ocean, while at the high range

Dome could be the owner of fabulous riches. Its share-holders still held a sweepstake ticket.

Sometimes Dome's ability to pluck figures out of the air is little short of magical. On the sole basis of its Kopanoar discovery well eighty-five miles northwest of Tuktoyaktuk, and *before* it had completed the necessary stepout wells to determine whether the discovery had any commercial value at all, it made a detailed calculation showing that the estimated development cost of the Kopanoar structure would be $15,000 per daily barrel of oil, which was about twice the cost of the Forties field in the North Sea but significantly less than the $37,700 cost of the Athabasca tar sands. Jack Gallagher presented this information to the Toronto Board of Trade in an address in January 1981. In the same speech, he said many of the structures in which Dome had an interest were individually as large in area as the Prudhoe Bay oil field off the North Slope of Alaska, where in excess of fourteen billion barrels of oil had been proven. If Dome's success ratio of oil discoveries to wells drilled was maintained, this could mean a small Middle East in the Mackenzie Delta! When Gallagher employs this kind of rhetoric, so outrageously optimistic, it is small wonder that people cannot decide whether he and the company he is promoting are endowed with the counterfeit qualities of a three-dollar bill or, as one Calgary oilman put it, whether they are listening to John Kennedy promising that he was going to put a man on the moon. In holding out this kind of hope, is Gallagher a charlatan making empty promises, or a visionary cast from the mold of those great nation-builders a century earlier who, in the face of incredible skepticism, promoted the "impossible" Canadian Pacific Railway? Gallagher may yet earn a noble place in Canadian history books, but many people who have invested in Dome on the strength of his optimism will not know whether their faith was justified for quite some time.

Reflecting early in 1983 on Dome's unrestrained pre-dictions, Gordon Harrison said, "I guess we did have a tendency to be somewhat promotional to get the Beaufort to be considered a real possibility by Canadians. In the days when we were first starting, it was very

difficult to get people to take it seriously. My colleagues
[in other companies] would stop me on the street and
pooh-pooh the whole idea of getting oil from the Arctic.
There was a time when our competitors used to tell
people privately that production in the 1980s was absolutely absurd. Even Gulf will admit it's possible in
the late eighties now. But I do agree that it [Dome's
propensity for making ambitious claims] has probably
caused some credibility problems for the company as
well."

Talk of "another Prudhoe Bay" or "a second Middle
East" certainly invested Beaufort drilling with an exciting sense of drama (until listeners became fatigued
by frequent repetition), but in a sense such claims are
irrelevant. In deciding whether it can develop an oil
field commercially, Dome need only confirm an oil pool
of perhaps 500 million barrels—no mean accomplishment to be sure, but a far cry from the 40 billion figure
it has touted for the entire Beaufort region.

By 1982 Dome was thrusting ahead with its plans
to extract oil commercially from the Beaufort, and it
talked confidently of first production on a small scale
by 1986, although by year-end it began thinking more
realistically of 1987 or 1988. By the early 1980s, welcome circumstantial evidence was accumulating that
the region could indeed contain major oil fields, and
even if Dome's credibility in the witness-box was under
attack, no one could seriously challenge the impeccable
credentials of two upright citizens who emerged to stand
by its side: Esso Resources Canada Ltd. (the exploration
arm of Imperial Oil) and Gulf Resources Canada Ltd.,
tight-lipped, cold-eyed organizations, certainly not given
to rash stock promotion. Both had been quietly active
in the Mackenzie Delta–Beaufort region for a number
of years but, prodded by the federal government, they
announced very substantial new spending plans in the
wake of the National Energy Program.

As pulses began to quicken within the industry at
the prospect of oil and gas production from the Beaufort,
a serious difference of opinion arose between Dome and
Esso over the method to be used to transport Beaufort
production to southern markets, with the third major
player, Gulf, standing on the sidelines still making up

its mind. Esso believed a pipeline down the Mackenzie Valley was the best choice economically, and felt it had a good chance of obtaining permission to build such a line from the federal government, despite the much publicized Berger Royal Commission, which had studied a previous proposal to build a gas line down the valley and had concluded in 1977 that there should be a ten-year moratorium on any such scheme to allow for the settlement of native land claims.

Dome believed that it would be far cheaper, in the early years at least, to move Beaufort oil to market in huge icebreaking supertankers that would eventually cruise the Northwest Passage year-round. In Dome's view, the threshold of reserves needed to move oil by tankers was about a tenth of those needed to finance a pipeline, and the economics of tanker operations would remain more attractive than the pipeline alternative until the Beaufort was producing about 750,000 barrels of oil daily. The implication of this is clear: using tankers, Dome could begin marketing oil from relatively small fields, rather than waiting perhaps four or five years longer until huge reserves needed to justify a pipeline could be proven. For a company desperately needing to justify its faith in the Arctic, and wanting a quick financial fix, Arctic oil production could not begin a moment too soon.

In the fall of 1982, Dome, together with Esso and Gulf, published at a cost of $14 million a formidable nine-volume study known as an *Environmental Impact Statement*, which was to be the basis for a lengthy series of public hearings at which the leading proponents of oil production from the Beaufort Sea–Mackenzie Delta region were to be interrogated over the possible harmful effects of hydrocarbon extraction from the Arctic upon the environment and upon the people who live there. The study, almost six inches thick and two years in preparation, contained several hundred thousand words, maps, charts, photographs, graphs, and statements of persuasive reassurance that industry's purpose could be accomplished safely with the minimum deleterious effects.

The process of environmental scrutiny was just one aspect of regulation that Dome and its industry peers would have to undergo before they could obtain cabinet permission to move from the exploratory to the production stage in the Beaufort. By this time, more than $2 billion had been spent over two decades by the oil industry in searching for hydrocarbons in the Beaufort-Mackenzie region, and about fifty companies had exploration permits. The industry, asked to indulge in some highly speculative crystal-ball-gazing to assist in the government's planning, produced a scenario of what the Beaufort oil patch could be like by the turn of the century. The results were awesome: Beaufort oil production could amount to 1.25 million barrels a day and as many as twenty-six tankers, doubled-hulled, icebreaking leviathans, could be carrying oil to market. In addition, huge volumes of natural gas could be sent flowing south by pipeline or converted into liquefied natural gas and transported year-round in a fleet of LNG tankers. The Beaufort could be dotted with artificial islands and oil would be loaded onto tankers from huge man-made loading atolls. The ocean bed would be crossed with underwater pipelines, and a big fleet of drill ships, dredges, and various support vessels could be plying the frigid Arctic waters in every month of the year. There would likely be more than fifteen thousand people working in the Western Arctic oil patch, and the town of Inuvik on the Mackenzie Delta, the main supply and service center for the region, which has a present population of about three thousand, could see its numbers grow to perhaps twenty-one thousand by the year 2000.

Dome maintains that the Beaufort, in combination with other frontier areas, holds the keys to energy self-sufficiency for Canada. If production does not go ahead in the Beaufort and other exploration regions of Canada, the country could be importing as much as one million barrels of oil daily by the year 1990. Assuming a world oil price of $40 a barrel, this would result in a drain on the Canadian economy of $14.6 billion a year. Dome's solution is simple: spend that money in Canada and just see what the rippling, multiplier effect will be throughout the Canadian economy, benefiting all sectors of developed industry and diminishing regional economic disparities. The federal government would

The chairman of Dome flashes the famous Jack Gallagher smile.

A page from the St. John's (Winnipeg) High School magazine in 1933. Gallagher stands apart in the top right corner.

The young Gallagher in 1937, working for the Geological Survey of Canada north of Great Slave Lake in the Northwest Territories. On the right is his geological assistant, Steven Cosborn.

THE DOME EMPIRE

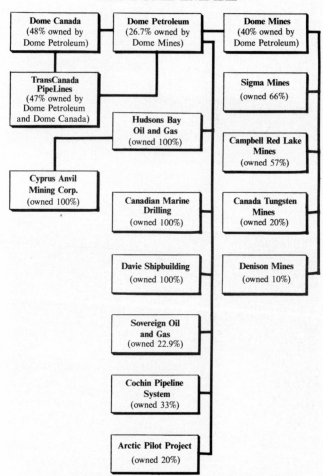

Dome's enormous interests as it fought to avert bankruptcy in September, 1982

Bill Richards, Dome's president and boardroom pugilist, announces in 1981 that Dome has bought Davie Shipbuilding for $38.6 million.

Gordon Harrison, who was head of Dome's Canadian Marine Drilling Limited (Canmar).

Energy Minister Marc Lalonde (right) tours Dome's drill ship *Explorer IV* in the Beaufort Sea with Gallagher in 1981.

Marc Lalonde celebrates the first anniversary of the National Energy Program in his Ottawa office in October 1981.

"I want the tooth fairy!"
September 1982, *The Calgary Herald.*

(TOM INNES)

July 1982, *Maclean's.*

(PHIL MALLETTE)

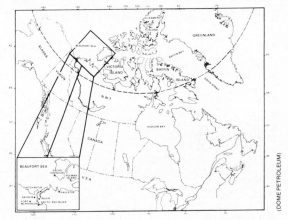

The Beaufort Sea, where Dome concentrates its search for oil in the Arctic.

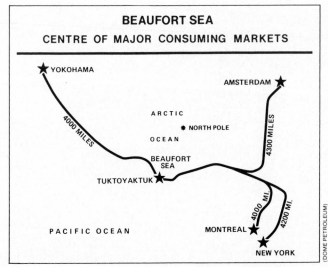

A potential center of ocean shipping, Tuktoyaktuk is almost equidistant from four major consumer markets.

Tuktoyaktuk as Homeland — the native hamlet.

(DEPARTMENT OF INDIAN AFFAIRS AND NORTHERN DEVELOPMENT)

Tuktoyaktuk as Frontier — Dome's northern base.

(DEPARTMENT OF INDIAN AFFAIRS AND NORTHERN DEVELOPMENT)

(AUTHOR PHOTO)

Inuit hunters from Tuktoyaktuk carve up a Beluga whale.

(CANAPRESS PHOTO SERVICE)

The New North: An igloo in the Northwest Territories becomes a garage for a precious snowmobile.

(DOME PETROLEUM)

Dome drill ships wintering close to the more traditional form of Arctic transportation.

(CANAPRESS PHOTO SERVICE)

A supply ship docks alongside Dome's drill ship *Explorer III* during operations in the Beaufort Sea.

(AUTHOR PHOTO)

Dome's drill ships and support vessels in their winter harbor at McKinley Bay in the Northwest Territories.

(AUTHOR PHOTO)

A Dome drill ship, followed by a supply vessel, breaks out of the ice at McKinley Bay at the start of the 1982 Arctic drilling season.

Crew change: Workers board a helicopter at the Tuktoyaktuk airstrip for a flight to a Dome drill ship.

For the uninitiated, the sight of the sea churning in a "moon pool" in the middle of a drill ship is unsettling. It is through this hole that the drilling equipment aboard Dome's *Explorer I* operates, sinking a shaft through the sea bed under the Beaufort Sea.

The powerful icebreaker *Kigoriak* breaks a pressure ridge, sending out ice cracks in all directions.

A cross between a ship and an island, Dome's huge semi-submersible drilling caisson (SSDC) was constructed from an old Japanese supertanker. It towers over a supply ship on the right.

The author stands in front of *Canmar Supplier*, which became firmly wedged in ice during his 1982 research trip. Eventually he was airlifted to Tuktoyaktuk by helicopter.

benefit through increased taxation, not to mention the 150,000 man-years of employment that could be created.

It is possible that during the 1980s, oil from the Beaufort Sea will power industries in Hamilton, fuel jets flying the Atlantic, heat apartments in Halifax, and perhaps even warm communal baths in Tokyo. But much is unclear. Dome's future in the Beaufort Sea depends on a complex and dynamic mixture of economics, politics, and technology. The attractiveness of expensive frontier oil production in Canada is linked crucially to the world price of oil, and that is largely determined half a world away from the Arctic by men who wear burnooses and not parkas. If the volatile OPEC cartel implodes and world energy prices fall sharply as a consequence, this action will weigh heavily against the promotion of expensive frontier resources. There is another part of the equation, however. As the search for Arctic oil is intensified, the chances of a major discovery increase each year; if the Beaufort delivers a mighty bonanza, which it could, economies of scale will play an important part in reducing production and transportation costs to offset falling international prices. Arctic technology is leaping ahead, and with each refinement, each modification to equipment, each improved operating procedure, a few more cents a barrel are chipped off ultimate production costs.

The final conundrum is the attitude of the federal government, which ostensibly is dedicated to reestablishing Canada's self-sufficiency in energy. It is debatable at what cost Ottawa will be willing to support Beaufort (and other frontier) oil development. If Canadians can buy oil from foreign sources at a price significantly lower than that involved in frontier production, what premium would the politicians be willing to pay to guarantee self-sufficiency? The government can no more chance the injury it would do to Canadian exporters by forcing upon them energy costs not borne by their foreign competitors than it can afford the displeasure of the electorate by making them pay domestic petroleum prices set artificially high.

The cost, the ambition, the sheer sophistication of the technological onslaught on Canada's North is breathtaking. Planned with computers, bankrolled by tax shelters and government grants, the southern assault on the Beaufort commands the best in brains, engineering expertise, and technical innovation. The Beaufort almost certainly contains riches that Canada will value highly one day, but whether that wealth will be exploited by our generation is still an open question.

Chapter Ten

ICE: THE FEARFUL ENEMY

Dome is drilling for oil in one of the most dangerous areas of the world, the treacherous transition zone between the slowly rotating permanent polar ice pack that crowns the world like a vast irregular skullcap and the landfast ice that clings to the shore of the Beaufort Sea for all but a few summer months. It is an area of continual menace. Large ice floes that can mangle steel and snap drill pipes advance and recede at the caprice of nature, smashing into each other and creating formidable pressure ridges, then breaking and re-forming in eccentric patterns. The floes may be small or they may be frigid leviathans that can bear down on a drill ship with a leading edge miles across. Often they are easy to spot and can be deflected by icebreakers; occasionally, in thick fog, their surfaces camouflaged by slushy melt pools with waves lapping over their sides, they can sneak up, undetected by radar, and threaten imminent disaster. Shifting winds, for instance, can create an ice hazard where none previously existed. A large ice floe, even close to a ship, is harmless as long as winds are pushing it away from the vessel; conversely, even a small amount of ice can push a ship off-station if the winds are high. In the Arctic, the weather, subject always to sudden change, influences everyone's thinking and dominates all operational plans. Fogs drop swiftly like opaque curtains and winds veer capriciously.

Dome's ships have drilled for oil for several seasons without major mishap from ice, although there have been a number of near-misses when ice has actually snapped three-inch-thick anchor cables, forcing a panic retreat. The ice hazard is certainly more formidable than the company first anticipated when it began operations in the Beaufort in 1976. Not a great deal was known about ice movements until then, and Dome literally had to acquire intelligence about its enemy while on the battlefield.

In half a dozen seasons in the Beaufort, Dome spent millions of dollars on applied research and engineering, and probably acquired more practical knowledge about ice in the Arctic than anyone else in either government or private industry. By 1982 it owned the world's first commercial icebreaker, which, although of controversial design, was intended as a forerunner of Arctic supertankers; it had built an artificial island dotted with more than $1 million worth of sensors to measure ice forces; it had changed the damaged propellers of an icebreaker *below the ice* in the blackness of an Arctic winter when the surface temperature was 110 degrees below zero Fahrenheit (counting the wind chill factor); it had learned how to smash its way through ice with puny supply ships and to split or deflect moderate-sized floes; it had operated a floating dry deck in winter (with questionable success); it had built ice runways, ice islands, and ice roads; and it had conducted two seasons of ice experiments on a tiny rock island between Greenland and Ellesmere Island.

The Tuk weather office, run by Environment Canada from Dome's base camp, is said to be one of the best-equipped in Canada for its size: it receives satellite weather pictures and 12-to-48 hour weather prognoses based on a variety of data, including winds at an altitude of 60,000 feet; it gets facsimile weather maps; and it can tap into the government's weather computer in Edmonton. Additional data is also received from eleven automatic buoys that have been parachuted onto the ice and from a Soviet ice island that has been circulating in the Arctic for many years. Twice daily during drilling operations, if weather conditions allow, ice reconnaissance patrols are flown over the drill sites. Ice is photographed by space satellites, and sophisticated airborne radar systems are used to spot ice movements. On board the drill ships, computers are used to predict

ice behavior and to display the expected behavior on a
TV screen.

To describe the many shapes and sizes of ice, the
World Meteorological Organization has set out an impressive, if occasionally quaint, nomenclature. There
are ice bergs (such as sank the *Titanic*), which are not
to be confused with ice floes; there is floating ice, sea
ice, ice of land origin, lake ice, river ice, frazil ice, grease
ice, slush, shuga and niles (both dark and light), pan-
cake ice, ice cake, and floeberg. Ice comes in consolidated pack, close pack, and open pack. There is even
bergy water. Ice deforms by rafting or ridging and can
also form into hummocks.

One of the more sophisticated pieces of ice detection
equipment available to Dome is known as SLAR—Sideways Looking Aperture Radar. This useful and expensive tool is mounted on an aircraft and scrutinizes the
Beaufort Sea in 100-mile swaths by sending out a radar
cone 50 miles on each side of the plane. The image is
recorded on a continuous film which is developed on
board and then dropped from the aircraft as it flies over
Tuk base; within four hours the film is analyzed, made
into overlapping mosaics, mapped, colored for easy
interpretation, and sent on its way to alert drill ships
of potential hazards.

In the summer of 1982, Dome's researchers were especially proud of a fancy new piece of technology called
SIAM (Shipboard Ice Alert and Monitoring system) that
had been developed by consultants and had had some
limited testing the previous season. In 1982, SIAM, whose
computer, nourished by an increasing diet of ice information, produced a colorful graphics display, had been
installed on two drill ships so that a drilling supervisor
could glance at a screen and see the proximity of ice to
the vessel on which he was working.

Better still, given the computer's predictive ability,
he could call for a display showing the expected ice
picture in relation to his own vessel at some future time,
say twelve hours hence. The ship, shown at the center
of the screen, was encircled by a red ring representing
time; if there was no ice showing within this warning
circle, the drilling boss could feel reasonably sure he
would have time to complete the work he was planning,

assuming (a major assumption) that he felt confident with the computer's ability to foretell events.

The importance of the new SIAM system was that it gave the drill crews on board the ships added confidence to work in the face of approaching ice. Sometimes a particular operation might require one day of uninterrupted work or maybe three days. Previously, if ice was in a known location, alert measures were put into effect and the drill crew would automatically assume that ice could force a ship off location. With drill ships costing about $750,000 a day to operate, running for cover unnecessarily wasn't encouraged. Even if an operation that required three days was not feasible it was still possible to tackle a job that needed only twenty-four hours.

From all of its research Dome learned to treat ice with considerable respect, but it also gained confidence with experience; with increased understanding and more sophisticated equipment at its disposal, it was no longer necessary to run for cover at the first hint of trouble as excessive conservatism had dictated in eary years. Now it was able to get a more accurate measure of its enemy and a more subtle appreciation of the threat it faced, and could decide how to respond accordingly.

Taking risks, trying out fresh ideas, and operating at the threshold of technology is a point of pride among the people who work for Dome. But occasionally things go wrong. Trying the experiment of operating its newly acquired dry dock at McKinley Bay during the winter of 1981-82 when temperatures ranged from $-30°$ F. to $-40°$ F., Dome ran into serious difficulties. The company decided to employ two supply ships with their propellers churning for weeks to keep the ice broken around the dock; the result was the creation of gigantic ice boulders that congealed and strengthened, much as a broken bone acquires added strength as the fractured parts are fused together.

It was necessary to keep the ice broken around the dry dock because the experimental icebreaker *Kigoriak* was docked from December 20, 1981, until February 1, 1982, for a major overhaul to repair ice and accumulated corrosion damage. "We had to keep breaking the ice around the dock to get her out. If we let it freeze it would be doubtful she could have got off," said Terry Graham, assistant supervisor on the dry dock. "Sea water freezes at $-1.9°$ C. If you can get warm water to come up from the bottom, you might keep the ice away. We tried bubbling [by pumping air from a compressor] but it was not successful. In theory, the warmth of the bubbles and the mechanical action they created was supposed to clear the ice, but it didn't work, probably because there was insufficient temperature difference between the bubbles and the sea water."

On May 3, 1982, the men on the dry dock ran into severe difficulties as they tried to dock the work ship *Supplier IV*. The dry dock is like a monstrous table, with unusually fat legs, set upside down. Ballast water is pumped into the dock to submerge it so that the ship to be worked on can float in and rest on preset blocks; the ballast water is then vented and the dock will float to the surface again carrying its new burden.

The engineers and sailors were unaware of the massive volume of ice rubble, created by the churning props of their supply ships, that had slipped under the dock. As the dock tanks were flooded, the deck of the structure was slowly submerged to five meters, and about 8,000 to 10,000 tons of ice, displaced by the descending dock, bobbed to the surface like corks and crashed in on the deck with a thunderous roar, tossing aside huge support blocks and bending a steel stairway as a child might twist a paper clip. Some chunks of ice were about fifty feet long and twenty feet high. When the engineers tried to shift the ice with a front-end loader, they found that the equipment was too feeble for the task, and they had to send to Inuvik for an explosives expert who blasted the ice into manageable fragments, which were then shoved to one end of the dock in order to accommodate the supply ship at the other end.

"*Supplier IV* was in dry dock about four days. We went down again to get her out and get *Supplier III* in, and the same thing happened again," says Graham. "Three times we had to blast away ice from the deck of the dry dock."

The dry dock, which has a lift capacity of 30,000 deadweight tons, was built in Japan and cost $116 million, which included the price of its tow by ocean-going

tug to McKinley Bay. Several engineers grumbled that
Dome had bought an inferior "bargain basement" model
and that the ice difficulty probably would not have occurred if it had been equipped with high side walls like
a more conventional dry dock.

If you look at an oversized wall map and crane your
neck to see the very top of Canada where it snuggles
up against Greenland, and if the printing is large
enough, you just might be able to make out Kennedy
Channel at about 82° latitude and 67° longitude. But
you would never be able to see Hans Island, a tiny speck
of rock about a kilometer in diameter that sits smack
in the middle of the twenty-mile-wide body of water,
where it is constantly bombarded by gigantic ice floes
heading south.

Summer lasts about five days, they say, and the island, which is the tip of an underwater mountain, is
frequently shrouded in fog. It boasts a small fresh-water
lake but nothing else to distinguish it, apart from sheer
cliffs and enormous ice pile-ups that attracted the attention of a Canadian government survey team in the
1960s and once merited reference in a scientific paper.
For Dome it was perfect, offering a unique chance to
observe and measure the forces involved as ice floes
crumpled and disintegrated on impact, something they
would have to face when they built islands in the Beaufort. Laboratory tests are useful, up to a point, but ex-
trapolating data from small-scale experiments is too
simplistic and not nearly as valuable as watching the
real thing.

Dome's researchers spent two seasons, 1980 and 1981,
on Hans Island, watching ice collide with rock and running up a bill of about $1 million. Ben Danielewicz, a
young physicist with Dome, says, "We weren't able to
raise enough interest among other companies to help
fund a harebrained scheme like this, so we had to do
it on our own, and we sold the research results after
we came back, which helped recover part of our costs.
These guys at Dome like far-out ideas and they are
good in that they allow you to state your case. But
there've been a lot of harebrained ideas that haven't
got off the drawing board. You have to be able to explain

what you want to do and why it's important to the company, and you have to do it in a few paragraphs or the people who approve the ideas or the money will lose interest and you'll never get anything done. This was a pretty crazy idea but we convinced them on paper that it would work, so they gave us a chance. The second year it was much easier getting the money, even though we asked for twice as much then."

Getting to Hans Island wasn't easy. First there was a commercial flight from Calgary to Resolute Bay, and then a journey by Twin Otter to Hazen Lake at the northern tip of Ellesmere Island. From Ellesmere the last leg of the journey was by helicopter. Costs were astronomical: fuel, probably the most expensive item, costs $15 a gallon by the time it is flown up to the roof of the world. A Hercules transport plane flew 120 drums of fuel to an army base on Ellesmere Island, from where the drums were taken by Twin Otter about two hundred miles to a gravel bar just opposite Hans Island. Most of the fuel was consumed by the Twin Otter's own engines on its ferry trips.

The gravel bar was just marginally long enough for the plane to land, says Danielewicz; in fact, some of the pilots refused to touch down there. "One time they went off the end of the strip and up on to a hummock—they were just going too fast. We had to unload the plane and we all got underneath the tail and lifted it around to get the plane back on to the landing strip. It's risky going up there, and that's one of the reasons we're not going up there again. I think you're just tempting fate. It's important data, but it's not worth risking lives."

The Dome team spent three weeks on the island in 1980, and about a month in 1981. They had four time-lapse cameras going continuously during the trip to record ice ridging as the floes—some up to several miles in diameter and twenty-six feet thick—collided with the island. They also flew out to the ice floes to set up instrumentation packages prior to the collisions, and to probe and test the ice for depth and salinity.

"We know for sure that it's the first time ever that ice failure has been measured over a large area like this," says Danielewicz. "Usually the only ice failure data that we have are from laboratory tests; we usually work on samples that are maybe half a meter across, but never in the order of hundreds of meters we're looking at here. So this is really a very important breakthrough, and I think it puts us years ahead of other oil companies as far as design goes."

But the research project wasn't all work. While they were on the island, the weather turned bad and they had to go into semi-hibernation, since snow and fog prevented them from making observations. "A lot of the time was spent catching up on notes, reading, sleeping, and waiting for the weather to clear up a bit." Since they had not been able to take a bath for two weeks, they built an efficient makeshift sauna out of a tarpaulin and several Coleman stoves, and then splashed around in the partially frozen lake that provided their drinking water.

Dome's publicity makes much of the company's successes, but naturally minimizes or contrives to ignore its technical difficulties. A case in point is the building of Tarsiut, its first artificial island in the Arctic, sixty-five miles from Tuktoyaktuk, which was far more costly and infinitely more problematic than the company's polished publicity material would suggest. One of Dome's engineers took a series of dramatic photographs that reveal the half-built island being attacked by an intense Arctic storm, enormous waves lashing the manmade structure and imperilling the work of many months and the expenditure of tens of millions of dollars. Naturally, they never appeared in the company's annual report! Tarsiut was a brilliant engineering accomplishment and an expensive victory of human ingenuity over the forces of nature. But only just.

Island-building in the Arctic is not new: Esso built the first one in 1972, and by 1981 there were nineteen of them dotted around the Beaufort. All were constructed in shallow water and used what engineers call a sacrificial beach—long, gradual slopes against which waves can break, dissipating their energy. Dome, however, wanted to build an island in relatively deep water, about twenty-one meters, so that it could drill a delineation well near to an earlier oil discovery.

Tarsiut—*Nighthawk* in Inuit—represented a new

technique in Arctic island-building, and brought home to the company in dramatic fashion the risks associated with newfangled technology on that unforgiving frontier. The island was designed for Dome by Dutch engineer Hans van der Wal, an expert in marine construction and dredging who has built harbors around the world. Tarsiut was built to serve two purposes: it would establish a potential base on which year-round operations could take place, and it would create a real-world engineering prototype to help test designs for far larger islands, which might one day serve as production islands and loading terminals for supertankers.

The earlier artificial islands built in shallow water, close to shore, were simply constructed by dredging sand from the ocean floor and dumping it in large piles, which were then protected by hundreds of sandbags secured by protective netting. Their side slopes had an incline ratio of 1:15. Clearly it would be impossible to follow the same practice in twenty-one meters of water, since the volume of sand required and the transportation cost would be prohibitive. After many hours of calculation and computer tests, van der Wal decided on a daring innovation: he would build an island with steep side slopes of 1:5. Since slopes of 1:3 occur in nature, he felt that in theory it ought to be possible to lay sand at such a steep angle without risk of underwater slides. Laboratory tests in Vancouver confirmed his calculations, but the sand would have to be laid with demanding precision.

Van der Wal's plan was to create an underwater sand pile, called a berm, and then crown it with four huge hollow concrete structures called caissons. Like enormous empty shoe boxes, caissons measure 11 meters by 15 meters by 80 meters, weigh 5,300 tons each, and cost a total of $27 million to build. The caissons were built in Vancouver and then towed up the west coast of North America, through the Bering Strait, and into the Beaufort, where they were joined together to sit end-to-end like a huge square doughnut on the carefully laid sand pile. Once in place they were to be filled with sand and joined at the corners by four forty-ton steel doors. Additional sand would also be pumped inside the hole of the doughnut, completing the island.

Dome's management was persuaded that the new technique was feasible and committed the company to spend $60 million on the island's construction. That cost included $1 million for instrumentation built into the island to measure ice forces. In typical Dome fashion, everything proceeded at breakneck speed. Only nine months was supposed to expire from the signing of the contracts in January 1981 to the completion of the island. In fact, severe weather problems forced a month's delay, with near-disastrous results.

For the actual construction of the island, Dome recruited Bill Janson, another Dutch civil engineer, who had worked on dredging operations around the world. Van der Wal had known Janson years before, and had someone contact him in Djakarta, Indonesia, where Janson was Far East area manager for a Dutch dredging firm. Persuaded by van der Wal, Janson travelled to Calgary to talk to Dome. "It came out of the blue while I was sitting out there in the tropics. I had never heard of Dome or the Beaufort Sea or what they were doing. I had to look up on the map where the Beaufort and where Calgary were!" Captivated by the challenge of building islands in the Arctic, he took the job. "Hans wanted me to start work the next day. But I had a responsible job in Djakarta. I just could not have been happy with that. It was almost three months later, on February 1, 1981, before I came to Canada to work."

The Dutch seem to have cornered the market in dredging, certainly as far as Dome is concerned: to build Tarsiut the company chartered two large Dutch dredges, *Geopotes X* and *Hendrik Zanen*. Work on construction of the undersea sand pile started at the beginning of July 1981 and was completed by about August 10.

As the island-building progressed, the fickle Arctic turned menacing. "We had basically good weather during July and up to mid-August," says Janson. "Then we got a fairly strong northwest storm in mid-August. But we could not wait any longer; we placed the first caisson on September 11, and the last one about four days later. We then had to fill in the area enclosed by the caissons, and the caissons themselves, with sand."

The island builders believed they were within a day of finishing their work when an enormous storm hit without warning. "It was after lunch on a Sunday. The waves were colossal; they were probably three-meter waves out at sea, but when the water came towards the island the amplitude seemed to increase dramatically. We had to abandon the island and call all the peoole off and move two barges from the island for safety. About 3 P.M. we saw that the sand was eaten away from the top and we called for a helicopter to move three bulldozers into a safer spot."

The storm was so severe that it tore the forty-ton corner doors from their hinges. "We also lost as much as 75,000 cubic meters of sand, about three days' dredging," says Janson. When the weather calmed, workers began repairing the damage, but they were knocked off the island a second and a third time by recurring storms. "A number of pipes and steel pieces were lost overboard. There was one massive clump of ice on the island. It really was one devastating sight. We were wondering whether we would be able to get the island completed that year.

"After the third time—either October 4 or 5—it got warmer and there was a thaw. We then had to fish for those doors again. About October 12 we managed to get hold of the first door and put that in place. At this point we sent for an expert from Holland, who brought with him a special type of filter cloth which would help retain the sand." To protect the weakened corners of the island, the engineers decided to drive in steel piles and fill in behind them with sandbags. "When we had the corners sealed, we started pumping sand in—that was October 15. We handed the island over to the drilling people by October 22. By that time the ice was getting very hard. We could not have worked another day there. It is only a four-month season and we lost about a quarter of our working time."

During this tense time the island's designer, Hans van der Wal, circled around Tarsiut by helicopter, a dreadful sinking feeling in his stomach as he saw the work of months attacked by the Arctic storm. But when he landed on a nearby construction barge he forced himself to make a show of cheery confidence to encourage the dispirited construction crews. Secretly he

feared that the island's construction could be delayed
by a full year.

The next difficulty was to get a huge drilling rig onto
the island. The rig, one of the largest of its kind in
Canada, had been built in Edmonton, and the disman-
tled pieces had been moved in 210 truckloads to Hay
River, from where it was sent down the Mackenzie River
to the Arctic on barges. By the time the island was
completed, ice had become a severe problem. Shifting
ice pushed supply barges ten miles from the island, and
it was several days before they were recovered. Despite
these enormous difficulties the rig was finally assembled successfully, and the first well drilled from the
Tarsiut artificial island—known as N-44—was begun
on December 10.

Drilling activities in the Arctic are always con-
strained by Ottawa's insistence that the company must
have the ability to drill a relief well quickly in the event
of a well blowout. To provide that capability, they built
an ice island a short distance from Tarsiut by pumping
water onto ever-increasing thicknesses of ice until a
solid foundation was created.

Operating throughout the winter of 1981-82, when
temperatures dropped to as low as −38° F., the crews
on Tarsiut were plagued by ice problems, and also had
to keep a keen eye out for polar bears. The island was
guarded by Inuit bear monitors who kept up patrols
with high-powered rifles. One bear got up onto the island's helicopter deck but then retreated; another, more
aggressive and presumably hungry, was shot on the ice
island about one hundred meters from where the drill
crews were working and living.

Because of severe crowding on the island, the drillers
decided to store packages of chemicals on the ice. "Three
or four acres of ice broke away with our chemical packages and drifted about five miles away. We had to use
a helicopter to bring all these packages back to the
island again," said rig superintendent Fred Sevenko.
Tarsiut was surrounded by ice all winter. "Ice coming
at you forms a ridge about fifteen to twenty feet high
and about fifteen to twenty feet from the island," said
Sevenko. Tarsiut also began to run out of fuel, a critical
problem in the Arctic, because of the unexpected drain

from construction machinery. Its supply, supposed to last until July, had to be replenished by helicopter in early summer, an airlift that continued for about a month.

Personnel moving to and from Tarsiut travel by enormous helicopters, an uncomfortable form of transportation in winter, since passengers must wear a great deal of heavy clothing to protect against the deadly cold and then struggle into bright orange floater jackets and clamp ear protectors on their heads to deaden the sound of the whirling rotor blades. Even in summer, the trip is tiring, but there is no time to recuperate on the island. On a sunny day one July, a chopper landed on Tarsiut bringing in personnel for a new tour of duty just two hours before dinnertime. A young mess girl who had just arrived was pressed into service *instantly* to help prepare the evening meal; she dropped her bags where she was standing and hurriedly started setting tables, arranging butter and juices, all the while giving her colleagues a breathless appraisal of the Calgary Stampede which she had visited the previous day. (She had had a marvellous time! But the prices! And the crowds! And she got no sleep at all last night! And that helicopter was so dirty—this complaint underscored by indicating a grease stain on tight jeans, a touch too stylish for the Arctic.) As she bustled around the canteen, the chief steward, apologizing, asked her to go quickly to clean and change the bedding in a room just vacated, because a toolpusher, also newly arrived, was scheduled to work at midnight and wanted to catch a few hours' sleep first. She cheerfully obliged, still gossiping non-stop about the great time she had had "outside".

Initially, Dome was uncertain whether it would want to retain the island beyond the first winter's drilling. By January 1982, when there were indications of promising hydrocarbon structures, a decision was made to leave the rig on the island over the summer. However, it was obvious to the construction group that the island would have to be reinforced to survive future storms; the engineers were not confident that the corners, in particular, could survive another severe battering. Dome decided to raise the freeboard of the island by about two meters by using two-and-a-half-ton gabions—big bundles of rock encased in thick wire netting resembling chicken wire, although much thicker. The rock was carried by truck from Inuvik along a winter ice road to a spot halfway between Tuktoyaktuk and Tarsiut. Then the plan called for Dome's icebreaker *Kigoriak* to load the rocks and take them the last part of the journey to Tarsiut. But *Kigoriak* got stuck in ice about forty kilometers away and could not penetrate an unusually heavy ice ridge.

Racing against the clock to complete the work before the onset of summer storms, Dome decided on an incredibly costly solution: the rocks would be airlifted by helicopter for the last part of the journey! Special heavy-lift choppers were brought in from Vancouver-based Okanagan Helicopters Ltd. to do the job. In all, a thousand lifts were required. Also stranded on *Kigoriak* was essential equipment, including a backhoe, needed for positioning the gabions and driving the corner piles. A Sikorsky Sky-crane was brought in from the United States to move this material. The entire cost of the first phase of the repair work at Tarsiut was about $6 million, or ten per cent of the original cost. After this, the company still needed to spend another $7 million on raising the island even higher and adding rocks around the outside of the caissons to protect them from ice-scarring.

Dome officials say they are well satisfied with Tarsiut, their first attempt to build a deep-water island in the Arctic. Construction was extremely difficult and they overran their original cost estimates, but they never expected it to be easy. Dome, working at the leading edge of Arctic technology, is getting an education the hard way.

Chapter Eleven

A PRIVATE ARMADA

Man has put few vessels on the oceans of the world as ugly as the maritime abominations that serve as drill ships. Not even their designers could love them. The bridge superstructure is set so far forward that it practically overhangs the bow; at the stern a large helicopter pad is grafted on like an enormous dinner plate. Amidships, a 180-foot derrick towers over a "moon pool", a large square opening in the ship's bottom in which frigid dirty water sloshes around, and through which the wells are drilled. Every available inch on deck is packed with stores and equipment, the most obvious being large piles of drill pipe that are used to penetrate hydrocarbon reservoirs many thousands of feet below the ocean floor.

These floating contraptions scarcely seem entitled to the designation "ship". They move infrequently and travel short distances at that; they are, in fact, complex floating factories that thump and vibrate around the clock as their inhabitants follow a punishing twelve-hour-a-day, seven-days-a-week routine. Part of the reason for the frenetic activity is geographical; since this enormously expensive hardware spends half a year or more paralyzed in the frozen embrace of the Arctic ice, the ships must earn their keep by ceaseless toil in the brief summer months.

Starting from a relatively small operation in 1976, Dome has built the largest fleet in Canada, more than forty vessels of many shapes and sizes. The level of activity as the ships prepare to break out of the ice of their winter harbor at McKinley Bay on the Tuktoyaktuk Peninsula in early summer is breathtaking: helicopters buzz around like bluebottles at a picnic, and on those days when thick fogs do not prevent flying, there are about four hundred landings a day, as stores and personnel are ferried around the fleet. The armada now comprises four ice-reinforced drill ships, ten icebreaker-supply vessels, four support tugs, an ice-reinforced fuel tanker, about half a dozen barges, including an accommodation barge, three large ocean-going dredges (among the biggest in the world), the experimental icebreaker *Kigoriak* and the icebreaking supply boat *Robert LeMeur*, a big floating dry dock, and a supertanker bearing almost 100,000 cubic meters of oil to keep this thirsty fleet replenished with fuel.

One of the latest, and oddest-looking, additions to the Dome fleet is the semi-submersible drilling caisson (SSDC), which was introduced into the Arctic in 1982 to become its second man-made island, Uviluk. Dome bought an old 150,000-ton supertanker in Japan for just $5 million, cut it in two, extensively reinforced the forward section with additional concrete and steel, and then had it towed to McKinley Bay, where a large drilling rig and crew accommodation were installed. From the day Dome thought of the idea of the SSDC to the start of drilling on November 12, 1982, only eight months elapsed. The marine contraption they created is a cross between a ship and an island: a vessel that can be towed by tugs from one drilling location to the next and then, by ballasting, can be sunk onto a sand pile created for it by dredgers on the ocean floor. The importance of the SSDC is that it gives Dome the ability to drill year-round at a substantial reduction in costs, rather than being restricted to a brief summer season, as with drill ships. Yet, unlike the Tarsiut island, from which Dome also drills year-round, the SSDC is highly mobile; simply by venting its ballast tanks, it reverts from "island" to vessel. True to fashion, Dome executives were wonderfully enamored of their new system, which formed an island fortress that could apparently withstand the constant buffeting of winter ice. The marvellously fertile imaginations within the company were working overtime: if $5 million could buy a supertanker less than ten years old (there is a world-wide oversupply of such vessels), perhaps they could string together three or four of them and build a big production island from which icebreaking supertankers might one day be loaded with Beaufort oil.

Gordon Harrison, president of Canadian Marine Drilling Ltd., Dome's northern operating arm, says the company now has the technology to produce oil—if it is found in commercial quantities. "There is no use finding oil in 150 feet of water if you can't produce it; there is no use finding oil in the moving ice pack if you can't produce it. With the semi-submersible caisson we can drill ten wells, and if we have a [commercial] discovery we have that sitting out there to put us on production." He contrasts this with the situation at Hibernia off the east coast of Canada, where large reserves of oil have been confirmed but production technology is far from developed.

Dome insists that the Northwest Passage, for which European explorers searched for centuries, will one day become a major maritime artery linking the Pacific and the Atlantic oceans, and opening up Canada's Arctic mineral and hydrocarbon wealth to increasing world markets. Dome is planning a fleet of 200,000-ton tankers that will be able to navigate the passage all year. Artists' impressions of the proposed Arctic supertankers that Dome has on its drawing boards bear little resemblance to a conventional ship; in fact, with their super-streamlined design they seem more like futuristic high-speed railway trains. The proposed Class 10 icebreaking tanker (meaning it can break through ten feet of ice at three knots without stopping) would be a monstrous vessel, almost as large as the biggest supertankers now afloat. With a proposed length of 390 meters, it would be almost 100 meters longer than the passenger liner *Queen Elizabeth II*. Because the proposed Arctic vessel will have a double-sided shell and a double-bottom hull to minimize the risks of an oil spill in case of collision or grounding, it suffers a severe penalty in cargo capacity compared with the more conventional crude carriers.

Dome has studied tanker failures over a decade to produce what its engineers believe is the safest design possible for the Class 10. They claim that no single mistake, in design, operation, navigation, or mechanics, will lead to a disaster. Even if both hulls are penetrated in an accident the ship will have sufficient ballast to retain any spilled oil between the hulls. The ship will have many unusual features: the hull shape will incorporate an icebreaking bow with reamers to ensure that the channel cut in the ice is wider than the ship to allow greater maneuverability and reduced hull friction; the stern will be of special design to deflect ice away from the propellers; and because of the extra strengthening built into the vessel, it will be about three times stronger than a conventional ship. Dome believes that even if the tanker collides with an iceberg, it is very unlikely that the cargo tanks within the double hull would be ruptured. One company official observed that it could hit the iceberg that sank the *Titanic* and survive! Dome says its Arctic tanker will be able to pass unassisted through all ice in the Northwest Passage throughout the year. And as a result of design improvements gained through its research and development work, the company calculates that the ship will now only require a third of the horsepower it originally contemplated.

Dome has been talking about such vessels for more than five years and has built a couple of scaled-down prototypes, but the icebreaking supertankers have still not progressed beyond the company's drawing boards. To construct the massive fleet of Arctic icebreaking supertankers and other assorted vessels that Dome has in mind will demand immense industrial activity. If the ships are to be built in Canada, shipyards will have to be established specially to accommodate such construction, and existing facilities will need to be expanded. The prospect of such industrial spinoffs, promising greatly enhanced economic activity and many thousands of new jobs, delights both political and industrial leaders, and Dome has not hesitated to dangle an industrial carrot before any audience prepared to listen. It bought Davie Shipyard at Lauzon, Que., in 1981 with the intention of expanding so that the company would be poised to begin Arctic ship construction when needed, and it designed an entirely new Canadian shipyard which it had planned to build (at an unspecified location) for Beaufort development through a technical assistance agreement with the Japanese shipbuilding giant, Kawasaki Heavy Industries. Canadians were being promised the opportunity to become world leaders in icebreaker technology with the potential of exporting Arctic vessels to other countries.

It was a fond dream, but for the moment it has faded. Sadly, Dome's expansion plans for Davie have been shelved until the company can recover from the debilitating setback it suffered during the recession of 1982–83, saved only by a last-minute infusion of revenue when it gained a contract to construct a ferry for Canadian National Marine. The chances of Dome constructing a new Canadian shipyard also seem far less likely as the company struggles to survive its massive debts, directing such capital spending as it can afford to projects offering a quick return. Long-term industrial strategy, so sensible and so appealing, must await better days, and if supertankers are built for the Arctic, the likelihood is that they will be built in Yokohama rather than in Halifax or Vancouver.

Politicians and moralists who bemoan Canada's lost work ethic should witness a crew-change on board a drill ship for an experience they would find truly inspirational: the comforting assurance that on the frontier at least Horatio Alger still lives! Fresh workers—or, more correctly, replacement workers, for they have already spent many hours travelling, mostly from distant southern cities, and have often wilted by the time they arrive in the Beaufort—must expect to begin work immediately if required, since the crews whose duties they are taking pass them, suitcases in hand, as they descend from the helicopter deck, and occupy the seats on the chopper that the new arrivals relinquished only moments earlier. There is no time to settle into new living quarters, to ease into the job gradually—work begins at once and at a full and hectic pace.

Despite the advances in technology and the invasion of a handful of women onto the drill ships, it is still a "macho" world: muscle-punishing, sweaty, loud with cussing, and dangerous. On the drill floor, mud-bespattered roughnecks use giant tongues to wrestle

sections of drill stem into position, all the while alert
for evilly whirling chains that whip around like a knout
and can easily amputate a finger, or worse. But today's
oil rig workers live very differently from their predecessors several decades ago. In the old days, they were
barely educated farm boys with no fear of physical labor, who endured poor living conditions, disgusting food,
the bullying of toolpushers (foremen), and many a bru-
tish prank by fellow workers, to emerge eventually as
seasoned oilmen. The route to promotion is still essentially the same today, but the industry now offers
training courses, and on the drill ships workers live
comfortably and dine with distinction. Remarkably,
Dome's workers in the Arctic eat fresher fruit and veg-
etables than many people in southern Canada. During
the drilling season the company moves 35,000-40,000
pounds of fresh perishable food *a week* into the Beaufort
to feed 1,200 to 1,400 people. Refrigerated trucks travelling via Vancouver and Whitehorse in the Yukon take
about seven days from the orchards of California to
Inuvik in the Mackenzie Delta, where they are loaded
onto a waiting barge for a twelve-hour journey down
the Mackenzie River to Tuktoyaktuk. From here the
fresh produce is flown by helicopter to the ships. With
a well-coordinated transportation system, Dome was
landing fresh food from California at Tuktoyaktuk in
1982 for a freight cost of about thirty cents a pound.

After an exhausting twelve-hour shift there is little
time, and less opportunity, for leisure activities. After
their meals the crews can watch a movie on a video-
cassette, play pool or cards, or glance at the ubiquitous
skin magazines, and collapse into bed early for eight
or nine hours' sleep, to dream of the crew-change flight
that will whisk them away from such a mind-numbing
regime of endless work.

Without exception the crew members are well paid,
and they accumulate substantial nest-eggs during their
tours in the Arctic. The younger ones are there to save
a down payment on a house, or simply to buy a flashy
muscle car with which to impress their girl friends; the
older ones are putting their kids through university or
salting away sizeable sums to enjoy a prosperous retirement. They are prepared to endure the isolation,

the cold, the darkness, the long hours of travel to and from the Arctic patiently, if not cheerfully, because they know the rewards are far higher than anything they could earn in southern Canada. If they choose they can afford vacations in the Bahamas, skiing holidays in Aspen or Vale, or the home renovations they have been dreaming about. They endure a constant background of noise (some find they need earplugs to sleep), and they eat, work, and sleep in close proximity to their workmates for weeks on end and thus have to be adaptable and easy to get along with. This is no environment for shirkers or grousers or people with an overwhelming desire for privacy. If they are not team players, the chances are that they will not return for a second tour of duty.

There are two distinct disciplines evident within the Dome fleet: that of the oil industry, a land-based operation that has moved offshore; and that of the marine tradition. The drill ships combine an odd mixture of both, and there are occasional strains between the captain of the vessel and the drilling superintendent as to who is the boss. When the written rules don't provide an answer, it boils down to strength of personality, says a skipper. "In other words, who can shout the loudest!" On board the supply vessels, the tugs, the dredgers, and even the dry dock, there is no ambiguity. Unmistakably, workers here are seamen, true to the traditions of that profession and no other.

Visiting the icebreakers and supply ships of the Dome fleet is like stepping into an English pub—except for the not incidental absence of drink; most of the accents are from Cornwall, Lancashire, Yorkshire, Surrey, and Kent, and the conversation is about soccer, cricket, and British politics. Canada's lack of home-grown maritime expertise is sharply underscored: for its port captains, marine superintendents, skippers, mates, and engineers, Dome has been forced to rely heavily on experience imported from the North Sea oil fields.

Carl Jewell, the harbor master at McKinley Bay, is typical of the type of international experience Dome has recruited. After several years commanding small supply boats, he spent a particularly horrid winter on the North Sea. "We had wind speeds up to seventy-five

miles an hour and fifty to sixty-foot seas. I think in one
period when we were at sea for six weeks, we did sev-
enty-two hours' work servicing the rigs and the rest of
the time we were riding out gales. I got pretty fed up
with it." Two former colleagues had already left for
Canada and so he followed. He was offered the master's
position on one of the supply boats then fitting out in
Vancouver before the Dome fleet sailed for the Arctic,
but within a month he was asked to take a marine
superintendent's job.

"I looked after the storing and getting the ships ready
for sea, and I did the early pollution exercises for the
government where we had to demonstrate that we had
the ability to control oil spills. I joined on March 15,
1976, and we left about June 25 to sail around Point
Barrow, Alaska. I think from the marine point of view
it was probably one of the most interesting parts of the
whole job, actually getting a fleet in. There were two
drill ships, the *Canmar Carrier* (which was a 26,000-
ton stores vessel), three supply boats, and we were towing a barge. We had a little Canadian Coast Guard
icebreaker who was supposed to give us some help but
had to turn back to escort a research vessel—not part
of our fleet—that was badly damaged. I flew the ice
reconnaissance for the ships in a Twin Aztec and we
escorted them round and flew with them all the way to
the Arctic."

Rounding Point Barrow is a hazardous operation,
since usually there is only a very brief navigation "window" when leads in the ice permit the passage of ships.
Some years the "window" never opens, and many an
old whaling ship was crushed in the ice and swept off
towards Russia. "We got stuck off Point Barrow for ten
days," said Jewell. "The fleet started to drift to the
north and nobody could move. Luckily an American
icebreaker happened to come past, and he was doing
all right, so we used his assistance."

The next year Dome brought two more supply ships
up the West Coast and through the Bering Strait, and
again Jewell flew patrols to spot open leads in the ice.
There were unexpected hazards, he found. "The Eskimos down in Point Barrow are far from friendly; they
throw bricks through truck windows and things like

that. One time we were taking off from the airstrip down there and all of a sudden the pilot banked the aircraft just on takeoff and I thought, 'What the hell is going on now?' There was a rattle on one of the wings and the Eskimos had shot at us with a shotgun and peppered the wing. We went down and reported it to the state trooper in town and he said, 'So what? You are about the third one this week.' Aircraft, it seemed, are fair game." Some of the Inuit in Alaska are less than enthusiastic about the activities of the oil industry.

Jewell commutes across the Atlantic to an old farmhouse and two acres of land he is reluctant to part with on the Isle of Wight. His tours in the Arctic vary in length. "I suppose now, working seasonally, I can spend four or five months over there [in the United Kingdom]. In the early days, and I think it is still like that, if you worked for Dome you worked for it whole-heartedly, body and soul. It's a full-time business." He remembers that when he first began working for Dome, he worked for twenty months in the Arctic, returning to his farmhouse only once, for Christmas. He decided that was enough and now works only part of the year.

Clive Cunningham, master of the icebreaker *Kigoriak*, began his career as a sixteen-year-old apprentice. He has tramped all around the world: on tankers, in the Mediterranean wine trade, and on passenger ferries, and has serviced offshore oil rigs in the North Sea, off Brazil and Borneo, off the Canadian east coast, and, since 1975, in the Arctic. His non-stop patter and self-deprecating humor belie a lively intelligence and professional competence. Cunningham works a month in the Arctic and then spends a month's leave near Duncan on Vancouver Island, where he and his wife own a restaurant.

"I know one guy who does his time up here and then goes away to the Bahamas and plays golf for the rest of the year. There are three other guys, all mates in the company, who went and bought a yacht and sailed around the Caribbean all winter. And there are other suckers, like me, who spend their leaves doing the washing up. I work at home from 7 a.m. until 1 a.m. and I tell people I come North for a rest!" The routine may be different, but it certainly isn't restful. When he is icebreaking, Cunningham tends to sleep very little, although exhaustion eventually takes its toll. "After a couple of days of battering around, you just collapse."

Cunningham loves the Arctic. "I saw eight polar bears in three or four days in April. It was pupping time for the seals, which are in their dens under the snow so you can't see them, but you see the polar bears diving into the snow trying to grab the babies. I saw them make a few kills. And I saw a pod of one hundred whales last year all around a drill ship, just circling in and out. Snowy owls land on your deck and stay there for days at a time. I have come into Tuk with owls on our deck; we have loaded our cargo and they moved out of the way, but they won't let you touch them. And then when we sailed out, they stayed on the deck with us. I have tried to feed them everything: smoked salmon, lobster, sardines, bread, meat—they won't touch anything.

"We had an incident with a grizzly bear the first summer we were up here near Herschel Island. It was three o'clock in the morning and we were sitting around gossiping on the bridge, and I could see this black spot on the ice." There was a trapper living on the island, a reclusive character who had a large collection of tropical fish and acted as a caretaker for the Dome fleet when it wintered at Herschel after its first season in the North. Cunningham went over to alert the trapper so he could shoot the bear. "We rushed across to his cabin and I fell through the bloody ice. So I got changed into some of his dry gear and we went off on his Skidoo and toboggan to grab this grizzly. We were carrying a door on the toboggan so we could get across the cracks in the ice, just an ordinary house door. I am sitting on this toboggan, I have got a movie camera, and there are melt pools on the ice, and we are going through this lot and I am getting thoroughly soaked, and after about four miles we go through this deep pool, and since I am sitting on the door I literally float off. Bob, the trapper, carried on without me because he is still chasing this grizzly. He kept running right over the horizon, at least fifteen miles, and never came back for me. The only protection I have got is a bloody door which I am carrying across the Arctic ice. What would I do if something comes and attacks me? Close the door and lock it? So I carried this stupid door all the way back across the ice, and an hour and a half later I got back to his cabin. I was cold because I was still covered with the slush that had been kicked up with the Skidoo. I was wearing his gear and he is only half my size so I can't even do the jeans up, the shirt is open, and I can't do his jacket up, but it was a bright, sunny day. And I thought, 'Well, piss on this hunting, I'm not into this any more.' The trapper ran out of gas and so he had to walk back about nine miles. He took several shots at the grizzly but missed each time. He is a Scotsman, as wild as they come."

Alan Murphy is a young Irishman who looks like a leprechaun and works as a seaman on board a drill ship, although he has an M.A. in English and has been a teacher in both Montreal and Spain. The job suits him just fine; he is single and made about $42,000 in ten months in 1981. Then he blew almost all of it travelling, spending most of the winter at Cape Verde off the west coast of Africa. "I have a friend in Calgary, an Englishman and a university professor, who has been to one hundred countries, and he inspired me. I look on this now as a very good part-time job. I can make enough money in five or six months to go away and do other things that I enjoy." In 1982 he was paid a basic wage of $11 an hour, which was the lowest sailors' pay. Skilled workers make far more. "A head welder told me he earned $72,000 last year [most probably an accurate statement] and remember, we live well on board the ship and have no expenses and can wear any old clothes."

On board another drill ship was Peter Eaton, a young political science student from Carleton University in Ottawa, who was gathering material for an honors thesis on Canada's major energy mega-projects. He hoped to get a research job with a major energy company when he finished university and had set his eyes on Dome. He had tried repeatedly to get an interview with Jack Gallagher and said if he could get just ten minutes of his time he would persuade him to hire him. But Gallagher was proving elusive. A rather intense young man, Eaton was captivated by Gallagher's dream of

northern resource development. He was intellectually
curious and opinionated and found he could stimulate
little response among the crew of the drill ship when
he tried to discuss abstract political notions. Eaton was
employed as a messman, which he said was probably
the highest-paid unskilled work in the country. "I will
walk out of here with $10,000 at the end of the [1982]
summer. I will have worked sixty-two days, twelve hours
a day. I make a lot of money but my talents are wasted.
I would like to get on deck where there is a bit more
action but now I do the laundry," he said. He was an
avid reader of *The Financial Post*, since, even after paying his university fees, he would have money left over
for investment.

If Peter Eaton is at the bottom of the social ladder
on a drill ship, Jeff Adams, the 32-year-old rig superintendent, is at the top. Travelling around the world,
Adams worked his way up from a roughneck's job over
the space of fifteen years in the oil business. Drill ships
have a curiously divided command requiring good li-
aison between drilling crews and the captain. "If I want
to, I can pull rank," says Adams, "but they don't like
to think I can. Between the captain and myself we share
a hell of a lot of responsibility. If anything goes wrong,
it is not Gallagher or Richards or Harrison who will
take the flak but Joe Blow sitting up here." He earns
big money, but recalls that when a friend once tried to
introduce him to some of the Toronto Maple Leafs in a
beer parlor, he was snubbed by the hockey players.
"When I was overseas I was making more money than
any of them stuck-up Maple Leafs and I wasn't paying
taxes. It is a high-paying business overseas, but then
it is high-risk, too, and there is not that much security
in it."

The people on board Dome's vessels who can really talk
knowledgeably about insecurity are those hired to do
the dangerous pioneering work of diving in the frigid
Arctic waters. When Dome's brand new icebreaker *Kigoriak* reversed over a hidden anchor buoy in the Arctic
in November 1979, its propeller blades were so badly
damaged that they had to be replaced immediately. The
job was done by divers working under the ice in the

blackness of an Arctic winter when the surface temperature was 110 degrees below zero Fahrenheit (counting the wind chill factor). The repair work went on for more than two weeks with teams of divers operating in shifts around the clock. One of those who took part, John Kissinger, said it was "a nice job. It was really interesting. It was not a job which you could be told how to do, because no one had done it before. I believe the U.S. Navy once attempted it, but they couldn't do it."

Dome was faced with the problem of repairing the pride of its Arctic fleet—in service just a few months—without the use of a dry dock, which it did not acquire until later. To facilitate the repairs, *Kigoriak* was taken about two hundred miles to Summers Harbour, a fine natural bay, which has excellent underwater visibility, an important consideration for the work to be tackled. A West Coast firm of diving specialists, Can-Dive Services Ltd., of North Vancouver, flew up a team of twelve divers from Vancouver and they began work on December 6. Temperatures often ranged from $-40°$ to $-60°$ F., with strong winds, and chain saws had to be used to cut through six feet of ice before the divers could descend. It was horrendously uncomfortable: often someone would be standing in four feet of water still cutting more ice. As the divers climbed out of the hole, their gloves froze onto the ladder. And the work demanded long dives, the average being close to two hours.

The men in the water, however, were warmer than their colleagues providing support on the surface, since sea water freezes at $-1.9°$ C. To protect those on the surface, shelters were built on the deck of *Kigoriak* and out on the ice. The operation was complicated when a tent caught fire from a heater and burned down, destroying all the diving gear: gases, hoses, and suits. Fortunately this mishap occurred as the men were warming themselves with a coffee break away from the shelters.

A television link was established to the surface so that engineers there could see what was happening underwater. Each of the four blades weighed about a ton and a half, and securing each blade were seven large fifty-pound bolts. Fortunately, *Kigoriak* had been designed so that the blades could be removed singly if the need arose. Sometimes there were two or three divers working together thirty feet below the surface, manhandling the heavy wrenches and bolts. The divers wore two or three pairs of long johns and several sweat shirts and pairs of socks, over which they pulled a dry suit and a helmet. When they had to carry down a special hydraulic wrench weighing 150 pounds, they added extra inflation to the dry suits for additional buoyancy.

The repair job successfully completed, the divers left the ship about the twentieth of December and arrived in Calgary on the twenty-third, gloomily contemplating Christmas in a hotel room, since the airlines would be fully booked. They needn't have worried: Dome, delighted with their efforts, had two Lear jets waiting to fly them back to the West Coast.

Divers, if not jacks of all trades, are certainly jacks of many. They need more than a smattering of knowledge about oceanography, chemistry, physics, and specialized medical emergency procedures; they are all mechanically inclined and usually have some experience, if not formal training, in welding. Since, like trapeze artists, they insist on preparing the equipment on which their lives depend, they also acquire skill in plumbing in order to test and repair their own life-support equipment. Divers are in increasing demand in the offshore oil industry, and so they need a good knowledge of drill rigs and the equipment employed on the ocean floor.

Darryl Rundquist, a diving supervisor working in the Beaufort, started scuba diving when he was nineteen, around Nanaimo, Ladysmith, and Chemainus on Vancouver Island, while working in a sawmill. "Every so often you would get a job picking up somebody's false teeth or something that they had dropped over the side of the boat and they would give you $5 to do it. I got tired of sawmills, tired of picking sawdust out of my ears and nose and everywhere else, so I took a three-month diving course in the U.S. that I saw advertised in a magazine. I learned how to use equipment and what it is like to work underwater and what it is like to work without seeing what you are working with; how to work by feel and touch. After a while it becomes more natural and you lose your fear of the water."

His first job was in the States for a couple of months after finishing diving school, but he was kicked out because he had no work permit. "I was nineteen and I didn't know any better," he said. He then went out to the east coast of Canada, where he found employment on semi-submersible rigs. "The first job I had was to recover a BOP [blowout preventer] that had been knocked over by an iceberg. That was really interesting and I really enjoyed it. I was not diving at the time, I was doing support work on the surface. One time on the east coast a seal tried to get into a diving bell with us. As we were going down, at about a hundred feet the fellow with me jumped back from the port and said, 'There's a shark out there!' I thought, 'There can't be a shark in these cold waters,' and I looked out and saw the tail fin of a seal." They got down to 320 feet and Rundquist left the bell through a hatch at the bottom. While one diver is out working, the other man stays in the bell with the hatch open, the air pressure in the bell keeping out the water below. "All of a sudden I heard this bang, crash, boom, smash, and it was the other fellow trying to get the wrench off the wall to hit the seal, who had stuck his nose into the bell. But it never bothered me."

Rundquist has worked in Spain, Singapore, and Taiwan, as well as on the Canadian east coast and in the Arctic. Like all divers, he has had one or two close calls. "One time when I first started, I was using scuba and I ran out of air. I came up and there was supposed to be a little Zodiac boat staying with our bubbles, but it was a bit rough and he got separated from us. I was being battered by this twelve or fifteen-foot sea against the caissons of the drilling rig. I tried to swim to the other side of the ship, which was two hundred to three hundred feet away, but I couldn't make it. I was just about to dump my tanks when some nice fellow happened to see me and threw me a line." His other close call was even more serious. "We use a combination of oxygen and helium in our saturation dives, and it is the helium, the inert gas that we don't use, that is absorbed into the tissues and creates bubbles in the bloodstream that worries us," says Rundquist. "You breathe gas at a deep depth, and as you ascend, these bubbles can't get out of your bloodstream fast enough, so they start to expand and you get the pain. It is *very* painful; I was paralyzed with it. I had it in the knees and they tried to get me out of the chamber three times. I couldn't walk and they started treating me with controlled doses of pure oxygen. I had so much CO_2 that I couldn't take any more. I was getting to the point where they would show me the mask and I started getting sick."

Many people who were professional divers for a few years have abandoned the work because they felt it was too hazardous. One of these was an English-born officer on the *Kigoriak*. Diving off the United Kingdom, he once get a version of the bends that could have resulted in brain damage—a nitrogen bubble in his bloodstream that caused an embolism in his neck. When he surfaced, his face was paralyzed, his head twisted to one side, his mouth open, and his tongue hanging out. Fortunately his boat men had the quick wit to replace his helmet and lower him back into the water to re-pressurize him. "I knew I was in trouble because I could not remember the boat man's name, although I had known him for five years. I stayed under water for about five hours, very cold and scared." He hit upon a novel mental exercise that saved him from serious injury. "What is one of the first things you learn at school? The Lord's Prayer. I could not remember it. I stayed down until I could remember the boat man's name and the Lord's Prayer and then I raised myself ten feet. If I could still remember them, I stayed there for twenty minutes and then raised myself another ten feet. When I forgot them again I would go down deeper for a while. That's the only time learning the Lord's Prayer was any use to me, but it certainly saved me that time." He suffered no permanent ill effects from his experience but quit diving soon after.

Experienced divers say that when the North Sea first opened up, the high pay attracted a lot of "green" divers just out of school who didn't know what they were getting into, and the mortality rate was high for a few years. But the record has improved because of more stringent safety regulations. Darryl Rundquist explains, "We have an emergency medical group that is on call—doctors who specialize in diving—and if anyone gets the bends, even a rash, we have to call them." In Canada, divers have a strict physical examination every year, and every two years they are all X-rayed to check for bone necrosis, a frightening disease often encountered by divers that can impair the mobility of the body joints.

Each Dome drill ship has a team of five divers on board during the drilling season. They are required to help relocate wells that were drilled but not completed the previous season (usually because of time constraints), and to install guide wires and blowout preventers; and they spend a lot of time checking that propellers and rudders are not frozen. The divers are on 24-hour standby, and can be called to work at any time of the day or night. Whenever the men are under water, their movements are monitored by colleagues at control consoles on the surface who watch rates of ascent and descent, pressures, gas mixtures, and even breathing rates. The average dive in the Beaufort Sea is about 190 feet, although some Can-Dive divers have gone down to 233 feet.

If the work to be done can be accomplished quickly, a technique known as a "bounce" dive is employed. The divers go underwater in a two-man diving bell which is capable of diving to a depth of 650 feet and is raised and lowered on a one-inch wire cable. An umbilical hose connection from ship to diving bell supplies the divers with gases, communication links, and power both for the bell and for their tools, as well as hot water that is circulated through their diving suits to maintain body temperature. Typically a bounce dive will last about sixty minutes, and afterwards the divers follow a decompression regime to reacquaint their bodies with surface pressures; for example, after an hour of working at 180 feet, the divers would require about three hours in a decompression chamber.

A saturation dive, referred to as a "sat dive", is more complicated and makes greater demands on the divers' endurance. It is employed when there is a requirement for extended or frequent diving sessions. A diver can be required to stay in "saturation" for several days or even weeks if there is much diving to be done. This means that the men, usually two of them, will be prisoners between dives, confined to a bleak pressurized steel chamber, thirty-two feet long and six feet in diameter, for days on end. They have bunks, a shower, a toilet, and a sink; music is piped in, and food, games, and books are passed in through an air lock. But it is a monotonous life; all they can do is sleep and talk and do a few sit-ups and push-ups and count their money. The money is significant: divers are paid on a bonus system depending on the deepest depth they reach during a particular dive. In 1982, Can-Dive was paying its men 75 cents a foot on a bounce dive, and $1 a foot on a saturation dive. So if they were put into "storage" at 160 feet—living within an atmosphere equivalent to 160-foot water depth—they would receive a bonus of $160 a day, in addition to their basic pay, for as long as the saturation dive lasted.

Although they don't think of themselves in a historic role, the men and women who work on the Arctic oil patch deserve to be called pioneers as much as did the homesteaders of the eighteenth and nineteenth centuries. Their work is arduous, frequently boring, and occasionally dangerous, although they live in comfort their forebears would never have imagined. Instead of ox wagon and plow they buzz around in helicopters, and employ computers, space satellites, and a bewildering array of sophisticated technology. They are attracted by high pay, a sense of adventure, and the prospect of career advancement, but their world is tough, cold, and isolated. Despite the disadvantages, if the oil industry's expectations are fulfilled, many thousands more workers from southern Canada will soon be commuting to the North to wrestle hydrocarbon wealth from beneath the depths of the frozen Beaufort—the new pioneers in virgin territory.

Chapter Twelve

TUKTOYAKTUK: THE TWO SOLITUDES

In southern Canada it was a joke before it gained an identity, an easy subject for the promoters of the T-shirt culture who covered bosoms with beer cans and simpering messages. By the late 1970s sweat shirts with the rib-nudging inscription "TUK U" were sported in Calgary and Edmonton bars by thirsty young men with big muscles and large pay checks just returned from the Arctic. To explain the joke, the words "University of Tuktoyaktuk" were added in smaller type. Intended or not, it was black humor of the grimmest kind, for there are few places in North America where education is received so apathetically and regarded with such poor esteem as in Tuktoyaktuk.

As the early Christians used to produce maps showing Jerusalem as the center of the known world, Dome has got into the habit of making similar maps showing the hamlet of Tuktoyaktuk (whose entire population could be housed on a couple of floors of a Toronto apartment block) at the center of ocean shipping. This is not such a farfetched concept, for, as a supertanker might sail, it is equidistant between Yokohama and Montreal (both 4,000 miles distant), while Amsterdam is but 4,300 miles and New York 4,200 miles away.

Tuktoyaktuk, or as it is usually called, Tuk, is really two worlds, separated physically and culturally, although the pathways between them are trodden increasingly. There is a small native hamlet with a population of about eight hundred Inuit and Métis, potholed, littered, and quaintly decrepit, occupying a small promontory surrounded on three sides by ocean; and about five miles away there is Dome's Arctic head-quarters, Tuk Base, with a bustling, noisy, no-nonsense ("Goddamn it, I don't care what Calgary says, we need that generator by Friday!") attitude to work, fortified by a nagging knowledge that drill ships, like surrealist taxis with meters ticking, are running up bills of about $2,000 *a minute*.

Tuk's most visible landmark is a legacy of the cold war three decades ago, an enormous DEW line (Distant Early Warning) radar station, one of a chain of elec-tronic sentinels that was erected across the tundra in the early 1950s to guard against intruding Russian bombers. The station, still manned, although obsoles-cent now in an age of spy satellites and supersonic rock-ets, was Tuk's *raison d'être*. Its construction offered cash wages to people who had depended before on hunting and trapping for their livelihoods, and its need for an airstrip opened up the Beaufort Sea to air transporta-tion. This coincided with an aroused awareness among southern politicians of native impoverishment owing to declining fur prices, and their commitment to provide, in modest measure, some of the essentials of a modern welfare society: a police post, a nursing station, gov-ernment housing, and a school. Thus prodded by well-intentioned authority, itinerant Inuit who had previ-ously camped near by on the eastern banks of the Mack-enzie River and traded furs and bought staples at the Hudson's Bay Company post settled into a new com-munity but tried to go about their lives much as they had always done, still travelling frequently to trap and hunt in the Mackenzie Delta, and to shoot polar bears and seals on the ice of the Beaufort Sea in winter and whales in the summer.

By the 1960s the oil industry's presence began to be noticed, and occasionally resented, as seismic crews dis-turbed trap lines in the Mackenzie Delta. But the major impact on Tuk began in 1976 with the arrival of Dome's Arctic drilling fleet. There was some friction that first season as Dome personnel, living on board a cargo ship in the harbor, occasionally whooped it up too enthusi-astically and fraternized too cozily in the hamlet. The local council, fearful of being overwhelmed and resisting assimilation, asked Dome to keep its distance. Sensing potential trouble, Dome backed away from the community, built its own camp—Tuk Base, limited access to the hamlet by its workers from the south to supervised tours and visits to The Bay store, and imposed a strict "no drugs, no booze" rule at its camp and on its ships.

One crude barometer of Dome's impact on the small native community is the price of bootleg alcohol: $120 for a forty-ounce bottle of hard liquor. Dome works its employees hard (twelve-hour days, seven days a week) and pays them handsomely, with the result that Tuk is awash in disposable income. Essentially, the company's pay scales are set to draw southern workers to an otherwise unattractive isolated posting, but since, correctly, it doesn't discriminate between workers hired in the north and in the south, the result has been a flood tide of money into a hamlet where few inhabitants have learned to handle a big pay check. It is not unusual for native workers who have had little previous experience with money to earn $30,000 to $40,000 in a six-month period, and the result has been a predictable aggravation of the traditional northern difficulties with drink and gambling, and the resultant disruption of family life.

Unlike some northern villages that have experienced severe alcohol problems, Tuk is not formally a "dry" community, but since there is no official liquor outlet in the hamlet, supplies have to be obtained from Inuvik, about one hundred miles to the south. It is legal to possess liquor for personal consumption, although resales are illegal. Police in Tuk believe there are probably five or six regular bootleggers in the hamlet. The liquor comes in by boat down the Mackenzie River in summer, or over an ice road or by air in winter. Getting leads on the bootleggers is not easy. "You have to get a guy almost red-handed. Sometimes an informer will let you know if they are pissed off because they did not get the last bottle," says an RCMP officer, who adds that even when the police do get evidence, the courts don't seem to look at the distress created by bootlegging. "In Fort Simpson, where I was before, we had a couple who were fined $1,000 or $1,500. One bootlegger told us he was making $3,000 on a weekend," says an officer.

Dome is acutely concerned about the potential use of liquor or drugs at its camp and on its ships. The luggage of all workers flying north on the company's aircraft is searched, and prominent notices threaten dismissal for anyone possessing liquor or drugs, a rule that even extends to employees of its subcontractors. Workers at Tuk Base talk of sudden room searches and even the use of dogs to sniff out drugs. Traffickers are constantly trying to pump drugs into the Dome camp, with uncertain success.

High-stakes gambling is also prevalent in Tuk, as in much of the North, although not apparently at Tuk Base. In the hamlet, an Inuit visitor from Sachs Harbour lost $6,000 in two nights of poker-playing. Games can last for up to three days, with some people playing continuously. There is also gambling for high stakes on board several ships of the Dome fleet, although it is no longer officially permitted. Still, one sailor admitted to losing $1,800 at poker. Before the official ban in 1979, games could involved $8,000 to $10,000 and frequently players went home without pay checks.

Roger Allen, Dome's supervisor of northern employee relations in Inuvik, says that poker is a favorite pastime for native people. "Within a community like Tuk or Coppermine, money changes hands so often there are really no big winners and losers. One day you may lose $1,000 and the next day you may pick it up again. But as long as money doesn't leave the communities, they don't seem to mind. When an outsider comes in and takes the money, then they become a little more defensive."

Dome tries hard to provide more wholesome off-duty diversions for its employees. In fact, it has pampered its Arctic workers to a degree that few Canadian companies have thought necessary. Its present prefabricated base camp, built in Calgary and brought up the Dempster Highway in early 1981 in 120 truckloads at a cost of $14 million, is superior by far to many hotels in southern cities. It can accommodate 364 people in comfort, and its facilities include a training center, a reading room, offices, bedrooms, and an excellent caf-

eteria, as well as a gymnasium, a racket-ball court, a
Jacuzzi, and saunas. It is constructed on stilts so that
the warmth from the building will not melt the permafrost and allow it to disappear into a quagmire of its
own creation. All buildings in the Far North are similarly protected or built on an insulating pad of thick
gravel.

The camp's food is famous, and, embarrassingly for
a company that spends so much public money, once
rated an honorable mention in a U.S. gourmet magazine. For a time the company went overboard, and its
Sunday buffet dinners were so lavish that they became
the talk of the tundra. The cooks, with few other diversions, would spend three or four days on the prep-
arations: live lobsters were flown in from Edmonton
wrapped in seaweed, and were displayed wriggling on
a stainless steel table so that diners could pick the ones
they fancied and have them cooked to order; the menu
would include cold glazed turkey, beef, and ham; min-
iature splashing fountains decorated the buffet table;
and sometimes there would be swans and animals carved
from ice. One special decorative feature was a B.C.
salmon mounted with a copper tube up its spine; cleverly arranged with a fly and hook and length of line,
it seemed that the curved fish was fighting an angler.
The salmon was glazed and the effect was further
heightened by the use of lights shining on different
layers of colored gelatin to give the effect of rippling
water.

One long-time northern employee recalled: "We got
visitors, bankers and guys from Japan, who couldn't
believe their eyes, and government people and politicians, who must have figured that we were living better
than they were. I guess for political reasons it had to
stop even before the company got into financial difficulties."

Still, even when cost-cutting was ordered as Dome
faced up to its financial crisis in 1982, no one was starv-
ing. By anyone's standards the meals were still generous, with thirteen to fifteen types of salads and eight
to ten different types of desserts each day. The entree
in a typical dinner menu could be chosen from roast
pork loin with apple sauce, or roast leg of lamb with

gravy and mint sauce, or beefaroni, or baked cod fillets
with béarnaise sauce, or mushroom omelette. Austerity
meant steak three times a week instead of daily, including breakfast. The dining room always had a big
table overflowing with fruit, including melon, grapes,
cherries, and plums, and there was milk, tea, coffee,
and cake available at all hours. Each television room,
furnished far more expensively than most employees'
living rooms, had a large refrigerator always filled with
apple, orange, and grapefruit juices.

There is a large reading room with deep armchairs
and sofas striped in rust, cream, and beige, coffee tables,
and reading lamps. There is a big stone fireplace, and
the knotty pine walls are hung with handsome Inuit
prints. On the days when Dome's jet is flying, the cur-
rent issues of the *Globe and Mail*, the *Edmonton Journal*, and the *Calgary Herald* are available in the reading
room. There are also numerous trade journals which
are largely ignored. The book shelves are filled with
westerns, science fiction, and trashy romances which
no one seems to read; in fact, there never seem to be
more than half a dozen people in the reading room,
which can accommodate twenty people in great comfort.

In another world, the hamlet of Tuktoyaktuk five miles
away, Vince Steen asserts that he dislikes the word
Métis and calls himself a "half-breed", which, he says,
means he has "the brains of an Eskimo and the gall of
a white man". It is a well-aimed jibe and he laughs
uproariously. An enormous Canada goose that he has
shot awaits plucking on his kitchen floor; his family's
evening meal comprises fish caught in a nearby lake
and Kraft macaroni dinner. He eats with his fingers,
attacking the fish head with particular relish.

As mayor of the Tuk hamlet, Steen is an outspoken,
even abrasive, critic of Dome, yet, like most people in
the hamlet, he benefits from the company's presence in
the community, and has a 55-foot crew boat which he
charters to them.

"I don't think the people realized what they had before the oil industry came in the mid-seventies. Most
of them realize well enough now, and they want to get
Dome and industry and the whole works out of town,"

he says. "They want to get away from the hustle and bustle and the noise of helicopters. There was a negative attitude to offshore drilling in the Beaufort, but people did not have a choice then about what was going to happen. We have a lot of people working on the base either directly or indirectly through contractors.

"I don't think people want them [Dome] to go, because they have adapted, but they don't want them to get any bigger. We have told the company, 'If you mean to get into ship-repairing, you should keep your workers out of the settlement.' Dome, like everybody else, have a hard time keeping control over their contractors. They might have a 'dry' camp already for their own employees but they can't impose it on contractors. For that kind of reason we asked them to take the heavy industry and ship-repairing somewhere else."

Dome has established its main harbor at McKinley Bay, about seventy miles to the east of Tuk, but the hamlet council has told the company it doesn't want to see a settlement created there. "We feel the guys are here for money and not here to live like us. If they start to bring in women and kids, it becomes a settlement, and we don't want that. They are already three times as many as we are. For the last two or three years, management have brought in their families; there are a maximum of six trailers for six families, and that is as much as we would allow."

Steen is a hunter and trapper, which means he must earn enough money to buy a new Skidoo each year, plus other equipment. A medium-sized vehicle cost $3,200 in 1982 and a fancy double-tracked Skidoo was worth $5,500. Used on rough ice while hunting polar bears, the less expensive vehicle lasts perhaps only two trips. As Steen talks, he slips a video-cassette into a machine and screens a tape he has made while out hunting, showing ice ridges thirty feet high, out in the sheer zone, a hazardous region of sudden ice movement created as the constantly rotating polar pack grinds against ice anchored to the land. The video machine is a new toy and the mayor obviously enjoys displaying it.

"That was the day we couldn't get the Skidoo to start," he says. "It was −35° F. We had a damn hard time that day. We had to take the engine apart." He gives a commentary as the camera pans along high ridges. "In that kind of stuff you go one mile ahead and five miles sideways. It takes hours and hours to get anywhere. You average about fifteen miles an hour on a Skiddo normally, but in rough ice like that you are lucky to average five miles an hour, or you will break your Skidoos or your sledge. Seals live in that rough ice. We hunt seals maybe once or twice a year for a different diet, but they are not really worth the effort cleaning."

Steen hunts polar bears on the ice about forty to sixty miles north of Tuk. "One year we were out on the sheer zone and it took five hours to get back. Only after we got a bear we started worrying about coming back home. The other fellow asked me, 'What are we going to do if it breaks up?' I said, 'We are going to learn Russian pretty fast.' There was nothing between us and Russia."

Hunters from the hamlet are allowed to take a total of twenty-six bears a year and each hunter is allowed only one tag, or permit, at a time. A good polar bear fur will fetch $2,400. "Out here it is serious hunting; you just live in a tent or a snow house when you are out in the sheer zone. A few guys make snow houses, usually to give extra shelter for the tents; that makes a double tent, which you heat with a primus stove. The average temperature is −30° F., but it can drop to −45° F. It is windy every day. Sometimes you are lucky and you hit a warm spell of −20° F."

When they are after polar bears, the hunters take dogs with them as an early warning system. "Some guys use them to stop a bear. The husky will bite the bear in the hind legs and make them sit down. The bear will make for high ground and try to get away from the dogs." The hunters will shoot when the bear is a little over a hundred feet away, but sometimes they may get as close as fifty feet.

Whenever Dome has people working on the ice it hires Inuit guards, known as bear monitors, to ensure the safety of its personnel. The company uses Jimmy Jacobson, a well-known carver from Tuktoyaktuk, as its agent to arrange the guards. An example of Jacobson's art, entitled *The Flying Eskimo*, is prominently displayed in the lobby at Tuk Base—Dome paid $3,500 for it in 1979, although it was the product of only three days' work. In the 1981-82 winter, Jacobson supplied eight men to work at the Tarsiut artificial island and at the McKinley Bay harbor sites. The guards are paid $145 a day, of which Jacobson retains $25 as an agent's fee. The oil companies have routinely engaged bear monitors since two Esso employees were killed by bears some years ago. Bears are harmless unless they are hungry, and then their appetites are voracious; of the two oilmen killed, only about thirty pounds of one man's body remained, and "the other man had no hips left."

The guards try to scare off the animals with the noise of their snowmobiles or by firing warning shots. But there is no way to discourage a determined bear that is feeling the pangs of hunger; he has to be shot, preferably in the vulnerable area near the ears. The bear monitors are not allowed to take the skins, which are confiscated within hours by the government; they merely receive a $35 fee for skinning the animals. What does the government do with the furs? "I don't know; give them to the Queen, I guess," says Jacobson. Although there were twenty bears around Tarsiut Island in the 1981-82 winter, only one had to be shot.

Jacobson has given up hunting, since he has to stay by the telephone in case a bear monitor is needed urgently. "You never know when they are going to call you. Sometimes they only give you half an hour's notice to find a man. They always take a guy when they fly out onto the ice in case they get stuck or fogged in."

Like all the older Inuit, Jacobson is intensely proud of the natives' culture and their ability to live off the land, emphasizing their superiority in local knowledge to that of the white man. Ben Danielewicz, a research scientist working for Dome, recalls how someone foolishly introduced him to William Nasogaluak, a former mayor of Tuk, as an ice expert. "It was stupid and very embarrassing because there is nothing that an Eskimo likes to hear less than that some guy from the south is an expert on ice. They are ice experts themselves. William kept ribbing me hard for about two hours. 'What kind of ice is that? When will this ice move?' I had to tell him, 'You know you know these things better than I do.'"

Although he cannot read or write, Jacobson says if

you show him a spot on a map seventy-five miles away
he can take you there in the darkness of midwinter.
Yet once in Edmonton he got lost about three blocks
from his hotel, and, seeing what he thought were a lot
of tough, mean-looking characters, he panicked and
called a taxi to take him back to safety.

Most of the non-native workers at Tuk Base have homes
and families thousands of miles away in the south, and
they live pampered bachelors' lives in camp while
counting the hours until their "crew change"—the day
when Dome's big blue-and-white Boeing 737 jet will
whisk them "out", away from the cold and isolation of
the alien Arctic that they get little chance to know or
understand. After their twelve-hour work shifts are over,
many change into fancy sports outfits and make the
empty hours pass by lifting weights, playing squash,
badminton, table tennis, or pool, or enjoying the whirl-
pool or sauna. If exercise doesn't appeal, they can read
that day's newspapers or watch TV in luxuriously furnished lounges. But John Nuttall's routine is different.
When his work day is done he jumps into his pickup
truck and within five minutes is with his family in a
cramped house only a stone's throw from the beach in
the hamlet of Tuktoyaktuk. Water is fetched in con-
tainers from a nearby lake and the family sewage is
taken out in plastic bags. A wooden box covered with
a tea towel with a picture of John Kennedy serves as
a coffee table. Nuttall, a slight man in his mid-forties
with a well-weathered face, and one of the few non-
natives living in the hamlet, believes he has the best
of both worlds. He successfully straddles two societies:
a noisy industrial world of high pay, demanding hours,
deadlines, and rigid discipline, and a tranquil world of
open tundra, forest, frozen lakes, and a silent solitude
only disturbed by an occasional rifle shot as he hunts
to feed his family.

Nuttall is a yard foreman at Dome's base camp. He
first came to the hamlet in 1972 with his Inuit wife
Elsie, whom he married in Yellowknife. At first it
seemed very bleak and isolated. "I remember coming
in in the first week of November. When I flew in, there
was a sort of a shock. I remember looking out of the

window of the plane and thinking, 'I won't be here long.' I am still here ten years later. Up here you can relax and enjoy life more." He works a regime of fourteen days of twelve-hour shifts followed by seven days' leave. Normally he takes the winter months off to indulge his love of hunting and trapping. "The people here have it made if they only realize it," he declares. "They can work for Dome and still go hunting and trapping. It is up to the foremen and supervisors; if we can spare the men we will let them go. This year most of the guys were given four or five weeks off for spring fishing and hunting."

Nuttall has become a conduit of understanding between hamlet and camp. He helps explain Dome and the imported southern work ethic to his native friends, and he interprets native aspirations to his white colleagues. He is loyal to Dome and believes the company has been good for Tuk. That does not stop him from fighting company actions he disagrees with; along with other members of the hunters and trappers association, he voted to stop a Dome plan to conduct oil-spill experiments at Tuk in 1982. "That's our drinking water," he says emphatically.

Despite Dome's conscientious efforts to be a good corporate citizen, the two worlds of Tuktoyaktuk—the traditional, slow-moving native hamlet and the intensely hard-working oil camp—coexist uneasily, and intercourse between them is fumbling and self-conscious. No matter how hard Dome tries to be a considerate guest, it remains a stranger. Sometimes resented for its effect on life in the North, occasionally welcomed for the material prosperity it has brought, it is always an outsider whose presence in the Beaufort is motivated solely by the prospect of oil. Once that is exhausted, in two, three, or six decades, the oilmen will depart and the natives will remain. For the Inuit and Métis of Tuktoyaktuk, the cardinal concern is this: what type of life will their children and their grandchildren inherit? Will the process of cultural assimilation speed up too quickly? Will the noise and pollution of the massive industrial onslaught occuring in the Arctic scare away the polar bears, and alter the migration of the beluga whales which the natives hunt each summer in the relatively warm, sheltered waters of Kugmallit Bay at the mouth of the Mackenzie Delta?

Dome and the other oil companies operating in the Arctic argue persuasively and with much expert testimony that they are a positive force in the area. But the native inhabitants of Tuk, enjoying a material prosperity undreamed of two decades ago, are confused and worried about the future as it approaches all too rapidly.

Chapter Thirteen

FRONTIER OR HOMELAND

Of all the challenges Dome faces in seeking to extract hydrocarbons from the Arctic, probably none is more sensitive than what is fashionably termed socio-economic relations, a vaporous, ill-defined field originally handled mostly by harried operations men who, by training and inclination, had scant awareness of social issues more complex than team selection at a curling club.

Dome has worked hard and long to make friends and influence the native people in the North. At first it was anything but welcome, and it is still often viewed with skepticism. The natives have reason to be worried: Dome and the other oil companies drilling in the Arctic are intruders who arrived, uninvited, in pursuit of their own enrichment and the resource interests of southern Canada, whose colony the North still remains. A persistent theme runs through all conversations: the crucial decisions that affect the lives and futures of northerners are taken in the bureaucratic enclaves of Ottawa, the oil company offices in Calgary, and the banking halls of Toronto, but rarely in Yellowknife or Whitehorse and *never* in Tuktoyaktuk. No matter how benignly Dome behaves in the Arctic—and most of the evidence suggests it has a lively conscience and an acute sense that corporate good manners are essential to its own self-interest—or how much money it injects into the Western Arctic economy, it remains an alien presence, to be tolerated and even profited from, but never embraced.

Gordon Harrison, who was president of Canadian Marine Drilling Ltd., the company's northern operations arm, recalls a disastrous visit to Tuktoyaktuk in the spring of 1976 just before its first drilling season. "I almost got run out of town. We put together a community meeting but it just didn't go well at all. People were very hostile and extremely vocal. One of the reasons was that the people there had been stirred up over a period of years, and they had had all sorts of people coming in and holding meetings, because of the Berger inquiry [into the Mackenzie Valley Pipeline]. They had been subjected to all sorts of viewpoints and propaganda and so forth from various [anti-development] sources. I must say they had some very genuine and sincere concerns as well; it was not all hype." (One's own information is never stigmatized as propaganda.)

After a few early blunders, Dome has, by most accounts, handled its relations with the people of the North well. But it had much to learn at first about the complexities of northern culture, values, and ambitions. The company is aware that it is operating in a goldfish bowl, its every move scrutinized by eager critics, prominent among whom is the CBC radio station in Inuvik. Harrison says, "I am sure you would not find any other project anywhere else in the world that has had the attention that this has had. People were simply looking for every little mistake you made of an operational nature—for example, insignificant oil spills—and trying to test us as to whether or not we would tell them, and we learned that we had better tell them. It was not very long before we learned that complete honesty and openness was the only policy to have. Oddly enough, that *is* something you have to learn. There is a bit of a tendency in all of us not to necessarily volunteer the bad stuff or the difficulties. We had a few stuttering starts in that respect, but once we really got committed to it, it worked out very well."

The priority that the company attaches to its relations with northerners is illustrated by a comment made by Susie Huskey, then the only native member of Inuvik town council. If she had concerns about Dome's activities, she said, she bypassed the company's local office, which she felt had no clout, and called directly to Calgary to talk to Harrison, who would invariably return her calls the same day, often within minutes. Few highly paid oil company executives would be that solicitous of a minor politician.

Harrison says, "When we got our drilling approvals, Judd Buchanan [Minister of Northern Affairs] told us that we must do something about the social matters of the north. We must bring social and economic benefits to the people there. I think it was interesting that it was given to us in that form; we were not given really specific directions. We were told we should give serious consideration and creative thought as to how we could have northern people participate in our programs and benefit from them."

After the disastrous Tuk community meeting, Dome decided it had to do something different, and it developed the idea of the Beaufort Sea Community Advisory Committee. "We wrote to seven of the communities around the Beaufort Sea and asked them if they would like to nominate a member to this committee. The committee would facilitate the information flow between us and the communities: their concerns about our operations and our description to them of what our activities were, and our outline of our problems and the answer to their concerns. They were pretty skeptical at first." But Harrison says the committee worked out well and is probably a prototype for other projects of this nature. Its members have visited Alaska and the Shetland Islands in the north of Scotland, to witness for themselves the impact of oil production in both places. Depending on who you are talking to, the committee is peopled by "Uncle Toms" who have been suckered by Dome with frequent expenses-paid junkets, or a tough-minded independent group of high integrity that has been successful in ameliorating the potentially harmful effects of development.

Susie Huskey says she joined the committee because she wanted to keep an eye on what Dome was doing. "For the first year and a half, the native committee members got a lot of flak from the peoples in the communities. We were traitors and we had been bought out by Dome, they said, but eventually it cooled down, and for the last couple of years [she left the committee in 1981], every time we went on a trip they didn't say too much and they listened to us.

"My greatest fear is an oil spill. My impression when we first met Gordon Harrison was that he was very abrupt and he said, 'We are going out to the Beaufort Sea to drill for oil whether you like it or not.' Of all the people sitting there, I said, 'That's fine, you can go out there, but you had better watch your oil spills because I am not going to eat whitefish bought from Alberta because your oil spills polluted ours.' I have been on their back for five years and my strongest concern was to see that the oil-spill equipment was constantly checked and the best in the world.

"I really forced the issue of hiring women; that was another big issue. They have quite a few women employees and that was because I really badgered them about it. In Alaska and the North Sea they don't hire women.

"In the first year, a lot of the native boys were not going back to work after their time out. The company asked, 'What can we do?' and we told them, 'Fire them. Tell them if they want a job they can come back next year.' Then we went to the communities to tell them of the problems that Canmar has with the turnover and the boys not coming back to work on time, just coming and going as they please. Word got around that Canmar would not tolerate this."

In 1981, the BSCAC had a budget of $120,000, all funded by Dome but administered by the committee itself. "We got criticized all the time that we were in their pocket. But when I was on the committee I would often go to a public meeting and criticize them to their faces. You keep track of all the promises that have been made, and when you go to a public meeting you do the follow-up right there. They get mad with you but that is too bad. The committee has managed to stay independent, but Gordon Harrison and I have had many battles." She revised her first impression of Harrison: "I found him to be a very shy, quiet type of person. He also sticks to his guns and I really think he is honest and tries to do the very best for people in the North. He walks around town in dirty old blue jeans and talks to anyone and remembers names."

The BSCAC was clearly set up by Dome as a way of circumventing the Committee for Original People's Entitlement (COPE), a political pressure group representing the 2,500 Inuvialuit or Eskimos of the Western Arctic. COPE refused to cooperate with Dome, since they have adopted the position that they are opposed to any development until their land claims are settled, but this did not stop them from supplying numerous services to Dome through a development subsidiary.

Much of the early socio-economic groundwork for Dome was handled by Mary Collins, who has a streak of stubborn determination to go along with some progressive ideas and a political science degree. After working as an assistant to the late Ontario Premier John Robarts, Collins had formed her own consulting company in Toronto in 1973, one of her first assignments being a review of social policy for the Royal Commission on Metropolitan Toronto. She became involved in the Arctic when she met Ernie Pallister, a Calgary consultant who was then advising Dome and other oil companies, at a conference in Toronto. Together they made a proposal for a communications program to familiarize the people of the North about offshore drilling. Initially, Collins worked for a group called the Arctic Petroleum Operators' Association, of which Dome was the biggest partner, and then for Dome itself. She spent an uncomfortable summer living in a small trailer in Tuktoyaktuk, ostracized by Dome's operational people, who thought she had been hired by the company brass to spy on them, and greeted suspiciously by the natives. "No one wanted me except the senior officers of the company. The operations people frustrated me in any way they could. They would not give me any cooperation with logistical kinds of things which are so important in the North. We laughed about this a couple of years later." She didn't laugh much at the time though, especially when, covered in dust as company trucks sped past her, she trudged about five miles from the hamlet to the base camp.

Collins says Jack Gallagher was far more interested in the work she was doing on the company's behalf than Bill Richards, who showed no sensitivity to issues such as native employment or the women's movement. But she remembers having lunch in Calgary with Harrison and a group of his operating subordinates, and being appalled by their chauvinist attitudes. "I wondered what on earth I was getting into. They were horrified at the thought of women on drill ships."

Collins quickly clashed with COPE, which had strong links with the local CBC station through Nellie Cournoyea, one of the group's leaders and a former broadcaster. "They were basically the adversary group, COPE and the CBC. We realized there was no way of working through COPE to get to the people, so we set up the Beaufort Sea Advisory Committee as an alternate group. COPE just hated that. I remember having a meeting with Sam Raddi [another native leader] and Nellie, and they just gave us hell. It was almost like a war between COPE and the company. It was one of those situations where you know early on you are not going to be able to find a solution."

Collins flew to the small communities around the Beaufort and chatted with anyone who would meet her. She spent an entire day in a small cabin waiting for a trapper to return home, drinking tea with his wife and listening to his children. On another occasion she flew to Sachs Harbour, which has a population of about three hundred at most, armed merely with a letter to the chairman of the local council, who, it turned out, had gone hunting. By the time she discovered this, the government plane she had travelled on had left and she could find no telephone. In another community she was told bluntly by a local councillor that she was unwelcome and should leave town immediately; she dug in her heels, however, and stayed for two days in the only hotel, establishing contacts in the community.

Slowly she developed a two-way communication line which involved reporting intelligence to senior Dome management and attempting to explain to the natives the benefits that Dome could bring to them. "There were lots of situations where I felt things were not being looked after properly," she said. In the fall of 1976 Collins wrote an extensive report for Dome which contained many recommendations about how it should implement programs to help northerners. The company must make specific commitments and live up to them, she argued persuasively.

Each year now the company negotiates with the federal government a socio-economic/environmental action plan which details the number of northerners the company will employ, its northern training programs, business development plans, environmental protection safeguards, and plans to protect the northern life-style and cultural heritage. Its performance is audited by Ottawa. Dome's involvement in the life of the North now transcends the type of activities that most companies would consider necessary. While undoubtedly motivated by enlightened self-interest and the business imperative, it tackles social issues with an awareness and a zest that is at times close to paternalism. It helps support a day care center and an alcohol treatment center, houses a technical training school at its Tuk base camp, and generally tries to make itself agreeable in a myriad of ways, such as the sponsorship of northern athletes at international tournaments. It even runs courses for its native women employees on home management and child care, extending to sermons on the need for a good breakfast for the kids!

A revealing anecdote is related by employees at Tuk Base. A former operations manager had arranged for a company helicopter to take several native youths fishing, but as the chopper was about to leave he received a call from a drill ship saying the aircraft was needed urgently to fly in spare parts. Here was a difficult management dilemma: which was more important—the spares or the effort to cultivate good relations with the local community? The natives got their fishing trip. Clearly such a decision would only have been made by a manager confident that he understood the policy priorities of the executives in the Dome Tower in Calgary.

Sometimes it seems to Dome officials that the native northerners have a far better appreciation of their operations than civil servants in far-off Ottawa who regulate their activities. Having made a detailed presentation to a very alert and skeptical audience one week in Tuktoyaktuk, they repeated the exercise a few days later in Ottawa. "We mentioned *Explorer I* would be leaving harbor late because of ice and they asked, 'What is *Explorer I*?' We had to explain that it was the

name of a drill ship and they said, 'What's a drill ship?
And don't your ships operate year-round? Could your
drill ships have an accident like the *Ocean Ranger* [the
huge drilling platform which sank in the Atlantic with
major loss of life]? Is it the same type of system? [It
isn't.]' We could not believe we were dealing with
such basic questions. After all, we were talking to a
director and six superintendents at the Department of
Indian and Northern Affairs responsible for northern
policies and land-use planning."

One of the most difficult problems for a southerner to
understand—and Dome was no exception at first—is
the natives' attitude towards work and their deep affinity with nature. A quotation from the 1977 report of
the Mackenzie Valley Pipeline Inquiry helps put the
issue into perspective. Mr. Justice Thomas Berger, who
heard more than a thousand native witnesses, wrote:

> To most white Canadians, hunting and trapping are
> not regarded as either economically viable or desir-
> able. The image that these activities bring to mind
> includes the attributes of ruggedness, skill and en-
> durance; but they are essentially regarded as irrel-
> evant to the important pursuits that distinguish the
> industrial way of life. This is an attitude that many
> white northerners hold in common with southerners.
> But the relationship of the northern native to the
> land is still the foundation of his own sense of iden-
> tity. It is on the land that he recovers a sense of who
> he is. Again and again I have been told of the sense
> of achievement that comes with hunting, trapping
> and fishing—with making a living from the land.
>   Much has been written about the capacity of the
> native people to wrest a living from the country in
> which they live. Only to the southerner does their
> land seem inhospitable; to the native people it offers
> a living. In every village of the Mackenzie Valley
> and the Western Arctic there are people who use,
> and feel they depend on, the land.

At first Dome couldn't convince native workers of
the need for punctuality in an intense industrial en-

terprise, nor did it appreciate the imperatives of the hunting seasons to which Berger refers. When the snow geese fly overhead or the beluga whales enter Kugmallit Bay, northerners employed in the Beaufort become restless and only with extreme difficulty sustain an interest in working as oil company storemen, drivers, or mechanics. At heart they are hunters still, not the amateur trophy-seeking marksmen of the south, but hunters who regard the activity with a deadly earnestness. They must lay in a store of food for their families, since meat produced thousands of miles away in southern Canada is prohibitively expensive and, in any case, is thought to be inferior in taste to the game available locally. Workers would simply disappear with the advent of the hunting seasons, much to the bewilderment and irritation of Dome's managers, who, unaware of the demands of northern culture, made no allowance for absenteeism. Also, the propensity of oil-rig foremen to bawl out native workers the same way they would routinely, and without malice, yell at workers in southern Alberta created unexpected tension: natives who live by consensus and mutual respect are unused to such treatment and are deeply humiliated by harsh words. Fearing chastisement, they would quit a job rather than report for work a few hours late, no matter how good their reason. However, in recent years Dome has made a great effort to explain to its native workers the need for punctuality, and the company also attempts to schedule leave periods to coincide with the various hunting seasons.

Despite earlier misunderstandings, Dome has had no difficulty in attracting native workers to its base camp and its Arctic fleet; in fact, it has often had to turn away potential employees attracted by the high wages it pays. For years the company was able to cream off the best workers available locally, although it now faces competition from Esso Resources, Gulf, and others. However, its high salaries have caused problems: it has to pay well to attract southern workers, and to pay its northern employees less would be discriminatory, but its high pay scales have disrupted local communities and small northern businesses which cannot compete with the wages offered.

Many northern workers still see wage employment as merely a necessary evil to facilitate the purchase of supplies that will equip them for their *real* life in the bush. With a few exceptions, of which the company is immensely proud (such as Cecil Hansen, the captain of its Boeing 737 jet), natives generally occupy the lower-paying and tedious entry-level jobs such as janitor, kitchen help, and yard laborer. This is not purposeful discrimination but a result of the priority placed on more traditional pursuits, and also of the appallingly low level of academic education in many northern villages. Illiteracy is high. The older natives had little opportunity for formal education and many younger ones don't care: since they can usually snap up high-paying work demanding a minimum of schooling and take home bigger pay checks than their teachers, they see scant incentive for higher education. The average education level around the Mackenzie Delta is grade six, and the Tuktoyaktuk school, which goes to grade nine, has an average attendance of only forty-five to fifty per cent. It is dangerous and simplistic to over-generalize, for there are a number of striking examples of well-educated and well-travelled natives, and slowly more are showing an interest in vocational training where it is available, but educational standards remain very low.

Father Robert LeMeur, an Oblate missionary who has served in the Arctic for thirty-seven years and after whom Dome has named an icebreaking supply vessel, says: "There are lots of children who do not attend school at all. Some attend school for two or three years and then quit. This is very bad, because it's the worst time not to have education. And there is a problem [of example], because some people who never went to school, and are bitter against school, still make a go of life and make a good standard of living, because they had a discipline in their blood and habits and they had to go through difficulties. So people say it is so easy: I don't have to go to school to get a job. Sure they get a job, but they don't get promotion. And this burns me, because in the Third World there are people who are crying to get education, because they want to take responsibility in their hands. So there is a contradiction for me.

They will never be capable of running their own affairs unless they have their own administrators, their own nurses, and their own teachers, lawyers, and police, and everything else. This is happening very slowly. You don't get too many people with high qualifications. Here we are passing through a difficult time because the kids don't want to go to school. They say, 'I can get money.' For me the most important thing is that they should get education and work for the oil companies or anyone else, because it is only when they get in there that they will get a strong voice."

One native person who was climbing the corporate ladder with Dome in 1982 was Roger Allen, a university-educated former Olympic skier, who was supervisor of northern employee relations for the company in Inuvik. He comes from an exceptionally athletic family: Allen was a member of the Canadian national cross-country ski team for seven years; two of his sisters were also team members, and one sister also went to the Canadian national curling championships, while a brother excelled in hockey. Allen won an athletic scholarship to the University of Colorado, and then worked as a floor hand on an oil rig and for both the federal and the territorial governments, where he became involved in the training of northerners. He was approached by Dome in 1979 to help handle northern recruitment and training. The sense of competition learned during his athletic career influences the way he approaches his job, he says. "I have been involved with so many different cultures now with so much travel, and through athletics I have seen how hard competition is necessary to make the national team. So I try to get a person who has a certain amount of ambition and give him the opportunity to develop that potential. I would like to see more native people become involved and provide the leadership in this type of organization. That's one area where we are lacking. Now that we have been involved with Dome for the last six or seven years, you don't see many native people in middle management or supervisory positions. I am in a position to advise the company and have had some impact. In 1981 we said that if production goes ahead, by 1985 we will need 6,500 employees in total, and we are trying to

maintain a twenty per cent northern complement, and
I think we have basically maximized the potential work
force in the Mackenzie Delta region and other regions
of the Northwest Territories.

"We pay our employees very handsomely. The average work season per employee last season [1981] was
eighty-nine days and each one averaged over $13,000
for about four months' work. A skilled-level person would
average $30,000 to $35,000 for a five- or six-month
working year. An exceptional employee can earn $60,000
in ten months."

Father LeMeur is saddened by the way the wage
economy has changed Tuktoyaktuk. "They have lots of
money. The mentality of the people has changed. They
are not as sharing as before, because they became more
selfish in a way. Before, we were helping each other
with various activities; now it is everybody for himself.
There are fewer visits now between ourselves, and less
cooperation too, because they get so much money and
they are more independent. In the old days a man was
proud that he was the best hunter so that he could give
to others. As long as there was food in a village nobody
would die of starvation. Because people are rich today
there is too much freedom in one way. There is lots of
money that is spent foolishly, that's for sure. They could
save more money, I think. Eskimo parents give $20
bills to young children to spend as they wish. But you
cannot criticize them or judge them with the mentality
of the south because here they have been in a hard
country. Sometimes they get in enough food to go all
year; the next year they get less and they have to go
through hell just to survive. Everything has been a
gamble. They go trapping and they may have five lousy
foxes or they may have fifty. They may have some caribou or they may have none. And the money is the same
thing. They use the money in the same way. It's a philosophy that's entirely different, which I don't necessarily approve of because they should not spend it
foolishly on what I call toys, whether electrical or mechanical."

Father LeMeur has coached his parishioners in ways
of dealing with the oil companies, teaching them how
to negotiate. And he is not above using political pres-

sure if he feels he has to. "I am dealing with the people here, talking to them, but mostly listening and acting as a counsellor; but I also have to be capable of meeting with those who make the rules. I tell them if the people think something is wrong. Talking to those guys [Dome and the other oil companies] you can have something better, because public opinion is very, very hot. They don't like to have criticism going to the paper or the radio or anywhere else, especially as we are in a minority around here. The people ask me what I think about things and I tell them. I have tried to give them the [negotiating] technique or the tactic that they should use when they need it. The people are capable of expressing their own opinions. They have been dealing with the oil companies since 1970, so they get wiser now."

Father LeMeur favors oil development but under strict rules, so that the northern people will be protected against oil spills and will benefit through better jobs and perhaps some cheaper goods. "We don't want everything to go north-south; we want something to stay in the country. We don't want to have a ghost town. When they extract the oil and ship it out, we want something to stay behind. There are lots of industries that could use the residue of the oil, like paint and plastics. Maybe tar or roofing materials; there is a possibility. I don't see why we should buy everything from Vancouver or Montreal if something could be done here."

The native people, so poorly equipped for the task academically, must confront an aggressive oil industry and resist, push, pester, and agitate to protect the lifestyle they want—and there may be confusion about what that is. Dome, of course, knows precisely what it wants, and its spear-carriers are well educated and articulate, knowledgeable about the North and at home in Ottawa's corridors of power.

Already the oil industry has brought great change to the Arctic. Its future impact could be immense—and no one knows unequivocally whether that impact will be benign or malignant. Those whose permanent home is in the North must live with the persistent nagging fear that one day a massive oil spill—from an oil-well blowout, a pipeline rupture, or a ship collision—may befoul great areas of the Beaufort Sea, creating a disaster of overwhelming magnitude that may ruin the life-style they cherish. Goodwill (or at least enlightened self-interest) and government regulation dictate that the oil industry will go to extraordinary lengths to ensure that every practical step is taken to avoid such a disaster. But even if the trauma of a big spill is avoided, there is still the possibility of chronic small seepages of oil as industrial activity is intensified in the Beaufort, and no one can say for certain what the cumulative effect of these would be over decades.

The native people also worry about the danger that icebreaking supertankers, crunching through many feet of ice, could pose to hunters pursuing polar bears or seals. Would they be stranded on the ice, their return to land cut off by the open-water track created by the ships? Dome has attempted to demonstrate by various experiments that the broken ice will congeal quickly in the wake of these vessels, but the hunters remain skeptical. Such questions will be exposed to widespread public scrutiny at government environmental hearings before the oil industry is allowed to pass from the exploration to the development stage in the Arctic. But the final decisions will be made thousands of miles away in Ottawa by politicians representing mainly populous southern cities, for it is with them that the ultimate stewardship of Canada's Arctic rests. It is they who will decide the future of the region that is to some a frontier and to others a homeland.

4

A NATIONAL CRISIS

Chapter Fourteen

THE HOUSE OF CARDS

Strong and proud in the Arctic and across the country, a symbol of Canadian entrepreneurship and daring until 1982, Dome was to find itself, in that difficult year, overwhelmed by the burden of debt it incurred to buy the Hudson's Bay Oil and Gas Company. For much of 1982 Dome lived a wretched hand-to-mouth existence, often days or even hours away from financial ruin; and it survived only through a nerve-wracking juggling act. It laid off staff, cut the pay of its remaining employees, sold everything that would attract a buyer, and delayed paying its bills. Its debts to creditors—those who supplied the many thousands of widgets and services that a large corporation needs to run its day-to-day affairs— amounted to about $1 billion, and if Dome had collapsed, the domino effect would have precipitated hundreds of bankruptcies in Western Canada. To cut costs, Dome sublet many thousands of feet of unneeded office space in Calgary at prices that helped depress the city's real estate market, and it sold off large inventories of drill pipe with the same result. Many suppliers tightened the screws; for example, a contractor that serviced the company's aircraft refused to release a jet engine until a $56,000 bill was paid in cash.

Trade creditors, however, were not the major problem: they could sue for their money if they wanted, and Dome was determined to keep them tied up in court for months, if not years. The real danger came from Dome's bankers, of which it has forty or more around the world; they could move summarily and call a default if a payment was missed. The company has cross-default provisions in its banking agreements, and so a default under any one becomes an automatic default under all of them.

"There is bound to be some wild man who is going to push it," said Bill Richards. And one European banker did, in fact, call its loans. "We had five days [to avoid default] and we stared them down several times on that," said Richards. "We were in a very difficult position. We felt we needed $600 million just to survive the summer, and we really couldn't see any way out of it. We used to say, 'Well, we can survive for another two weeks until such and such a day, and then God knows what will happen.' It was just fancy footwork that carried us through the whole of summer. We were making forecasts that showed that there was no way that we could survive. At times we were just hours away from going under. Sometimes we had one or two instances a *day* where we could have gone under and that's no exaggeration; not every day, but boy, we had some sweaty days. However, we never missed any payments without the consent of the creditors. The point is, once you go in default, the whole house of cards craters, so we have never been in default but we have been awfully bloody close."

What follows is Dome's diary of desperation from early 1981:

May 1981: As Dome plans its assault on the American oil and coal giant Conoco Inc., as a maneuver to seize 52.9 per cent control of Conoco's Canadian subsidiary, the Hudson's Bay Oil and Gas Company (detailed in Chapter 2), Jack Gallagher phones four Canadian bank chairmen: Richard Thomson of the Toronto-Dominion, Rowland Frazee of the Royal, Bill Mulholland of the Bank of Montreal, and Russell Harrison of the Commerce. Dome wants to borrow about $2 billion, he tells them in great secrecy, to finance the takeover. Can they attend to it for him? "Why, certainly, Jack, no problem. Call us any time." Within days all four banks are pressing ahead with a loan that they will bitterly regret. Significantly, the Bank of Nova Scotia, which does not share the same rosy view of Dome's prospects as its competitors, is not involved in the massive financing.

June 1981: Dome pays U.S.$1.68 billion for Conoco's controlling interest in HBOG. As security for the loans from the Big Four banks, which will have a ten-year term, Dome pledges the HBOG stock it acquires, plus some of its own oil wells as additional collateral.

September 1981: To reduce its staggering debt, a condition of its agreement with the Big Four, Dome puts its entire U.S. holdings, thought to be worth $1.5 billion, on the auction block. But it is a poor time to sell, and the best offer it gets is $350 million, a "fire sale" price which it rejects.

Fall of 1981: The Toronto-Dominion is worried about the size of its loans to Dome, estimated to be close to $700 million, and so its chairman, Richard Thomson, suggests to Dome that they sell the recently purchased shares and pay back the bank's money. Dome refuses.

Late October 1981: Dome has an embarrassing brush with the Bank of Commerce when it runs through its operating lines of credit and calculates that it will not be able to pay bills coming due about four days later. When it asks the Commerce for $150 million, Dome is told, "Well, we are not sure. We have an awful lot of money out to you guys." Jack Gallagher phones Russell Harrison and gets the money, which, as it turns out, is not needed after all because the company's receipts are higher than its payments. An insider recalls, "The banks were very critical of the way it was done. The place was really out of control. We were spending money like mad. Nobody knew when the bills were due. This incident put the banks on their guard and they said, 'These guys don't know what they are doing.' That was the first real sign that something was amiss."

Dome decides to buy the remaining forty-seven per cent of HBOG, for which it needs a further loan of U.S. $1.8 billion. Its Canadian bankers, already alarmed at their own exposure, with the Inspector General of Banks, Willian Kennett, beginning to drop hints of censure, refuse to plunge in deeper, and so Citibank of New York, which has long pursued Dome's business, assembles a consortium of twenty-six U.S. and foreign lenders to supply the money Dome needs. Dome's affairs, never straightforward, now become labyrinthine. The Citibank group, naturally, needs collateral for its loan and Dome proposes to give it HBOG's Canadian oil and gas wells. But there is a major snag: these are the major assets behind the HBOG shares already pledged as security for the previous loan to the Canadian Big Four. The Canadian lenders are horrified at the thought of losing their security, but they realize well enough how desperately Dome needs HBOG's cash flow, and so with considerable reluctance they permit Dome to assign the properties they hold as security to the Citibank group. The Citibank loan will take effect a few months later, on a fateful day in March when Dome's merger with HBOG is finally completed.

Late November 1981: The Big Four are becoming very apprehensive; having diluted their security, they put Dome on an extremely short leash. They plan to alter the term of their loan, also to take effect on that day in March, so that Dome must repay them $1.3 billion by September 30, 1982. If the company does not agree to this arrangement, the Big Four will refuse to allow Dome to complete the second half of the HBOG purchase. The enforced change in maturity starts a financial time bomb ticking in the Dome Tower, and it is this single event that precipitates the so-called Dome bail-out less than a year later. Yet no one within the company considers this short fuse to be a serious threat at first. When the new timing arrangement is made, Gallagher is actually out of the country and not fully aware of what is going on, although later he states that as chief executive he bears full responsibility for what happened. In addition to shortening the term of the loan, the Big Four chain down and padlock every unpledged Dome asset.

Early 1982: Dome has created an elaborate arrangement whereby it will simultaneously buy the remaining shares of HBOG and sell a chunk of HBOG's assets to others in the Dome family (such as Dome Canada and TransCanada PipeLines Ltd.). Canadian bankers are horrified to learn that not all the proceeds of these sales

will come to them as they had expected; Dome says it
has to give U.S. $400 million from the impending sales
to its American bankers.

In an attempt to reduce its alarming debts, Dome
officials have been scurrying around the world for
months trying to sell off assets. Not only are the company's own U.S. oil and gas properties on the block, but
Dome is seeking a buyer for the oil and gas acreages
of HBOG in Indonesia, Australia, Brazil, Egypt, Italy,
and the North Sea that will be its to sell once it gets
100 per cent control of HBOG. The Japanese are keen to
buy the Indonesian oil fields, since these are strategi-
cally close to home, but they don't want the rest of the
assets scattered around the world, and Dome, for whom
speed is vital, is reluctant to split the package and cause
further delays as it shops around for more than one
buyer.

Gallagher, ever the optimist, is promoting yet another ambitious plan, for a new company to be called
Dome International, which will help to avoid an outright sale of Dome assets. Dome will contribute its foreign holdings (including its extensive properties in the
U.S.) to the new corporation, and foreign investors will
inject a large infusion of cash. Dome will retain twenty
per cent interest in the foreign oil fields, which it will
manage; in addition it will walk away with about
$750 million in cash to help lift the heavy debt load off
its back. It sounds too good to be true. It is. Gallagher
tours international money markets for about three
weeks, falling in love with the Indonesian oil fields en
route, but nothing comes of his plans for Dome International. International financiers are candid—they can
buy oil and gas investments much cheaper elsewhere
in a depressed market. Some skeptics in the Dome Tower
believe Dome International never stood a chance, and
there are very quiet mutterings that Gallagher is out
of his depth. Richards, who steadfastly refuses to hint
at any criticism of his chairman, is horrified to discover
Gallagher is trying to involve Japanese investors in
financing Dome International at the same time that
Richards is trying to *sell* them the properties outright.
Not for the first time, it appears that the left hand is
unaware of the activities of the right.

March 9, 1982: A new actor is dragged kicking and screaming on stage. This is the Bank of Nova Scotia, which has previously shunned Dome's business. Scotia, however, had made a loan to HBOG, and since Dome will now be merging with HBOG, the Scotia loan becomes, in effect, a loan to Dome. Scotia is unhappy and nervous, and so it insists on exercising a right to the repayment of the considerable sum of U.S.$40 million (which has become due because some of the stock it holds as collateral has dipped in value). There is an unholy row between the Scotia and the other big Canadian banks, who say that the new actor in the Dome play enjoys better security than they do. But the men from the Bank of Nova Scotia insist on getting their money. Dome has been stripped of practically everything to which a mortgage can be attached and has nothing left to pledge as security to anyone, and so the TD, the Royal, the CIBC, and the Montreal, forced into the thankless task of lenders of last resort, reluctantly put up the U.S.$40 million demanded by the Scotia.

March 10, 1982: Dome formally merges with HBOG, and there are many legal documents to be signed at the Shearman and Sterling law offices in New York. (The paperwork is formidable, with signing ceremonies taking place not only in New York, but also in Chicago and Calgary, where Dome's Corporate Secretary, Harry Eisenhower, and Senior Vice-President, John Andriuk, sign doggedly on the company's behalf with aching hands from midnight until 7 a.m.) In New York the men from the Bank of Nova Scotia are still suffering from stage fright at the thought of any involvement with Dome, but they too must complete certain agreements. Before a large assembly of bankers and lawyers, a Scotia man is about to sign but gets cold feet and announces he has to call his head office in Toronto for authority. A conference call is arranged, to which the man's boss in Calgary is linked. Everyone listens to the drama as an amplified voice says: "I've already given you permission to sign." The worried banker signs his first name, then hesitates again. "Do you really want me to sign?" he pleads. "Just sign it!" comes the amplified order.

Of all the incidents in Dome's tangled and complex

affairs, the most remarkable involves the agreement of the Citibank group to provide U.S.$1.8 billion as backup financing to complete the second part of the HBOG deal. It seems that the money was advanced on a misunderstanding: the loan would never have been granted if the banking syndicate had known that the term of Dome's original loan had been switched, so that a payment of $1.3 billion in principal became due on September 30, 1982, just six months away! Since the two documents were dovetailed to fit each other as to the transfer of collateral, one can only wonder why *someone* in the Citibank group did not sneak a look at the Big Four's deal with Dome, which was being signed only steps away.

Bill Richards insists that the Citibank consortium *was* informed of this change in Dome's liabilities; members of the foreign banking syndicate say the first they heard of it was *after* they committed U.S.$1.8 billion in fresh financing. Members of the Citibank group say they relied on information given to them by Dome in November 1981, and on statements filed by Dome with the U.S. Securities and Exchange Commission. They did not truly understand Dome's changed obligations until Dome made a regulatory filing with the SEC at the end of May 1982.

A U.S. banker within the Citibank consortium asserts, "This second financing would never have been done if we had known of the change in maturity. It's very simple. There was no way Dome could have paid $1.3 billion in September. If we had known about the change in January, February, or early March, there would have been an instant reaction and the Citibank deal would have fallen apart. All the way through this, Dome were giving certificates saying there is no event of default, or event which with the passage of time would constitute an effective default. It's a judgement call, but they signed a certificate on March 10 saying there was no event that would constitute a default, and yet they knew damned well that September 30 was already cast in concrete. There is considerable resentment." Dome argued that it intended to sell sufficient assets to avoid a default, and that notwithstanding the September 30 date the Canadian

banks had said they would renegotiate the deal some-
how.

Late March 1982: By now, crippled by debt and with
its income dwindling alarmingly, like that of the rest
of the depressed oil industry, Dome is lurching towards
immediate bankruptcy. It cannot meet the $80 million
in interest and preferred dividend payments, and so
Richards asks the Big Four for the money to help pay
its March bills. The bankers are aghast and they refuse
adamantly. After all, it is only about two weeks since
Dome has taken on an additional U.S. $1.8 billion in
debt. When will the borrowing end? Richards' nerve is
still holding exceptionally well. That very night, as if
he has not a care in the world, the Dome president
throws a party at his ranch to celebrate the HBOG acquisition. Understandably miffed, the senior bankers
he has invited decide they will not risk getting manure
on their shoes, and they stay home.

March 30, 1982: At the dismal hour of 3 a.m., Richards,
Gallagher, the company's director of corporate affairs,
Colin Kenny, and several other executives fly to Toronto and book into the King Edward Hotel, to solicit
bankers from close quarters. Kenny is a useful man to
have around in a pinch. He is a political fixer with
impressive credentials, having worked before joini g
Dome in the Prime Minister's Office as Trudeau's li-
aison man with Ontario politicians. Liberal Party
workers did not enjoy his Prussian army habit of bark-
ing orders, and nicknamed him "Colonel Klinck". But
a Dome executive said later, "I could walk into his office
and ask him where Jean Chrétien is, and he would look
at his watch and say, 'In an hour he will be landing in
Winnipeg.' What's more, I'll bet he could get him on
the phone, too!" Kenny is to work overtime fixing appointments over the next few days. The Dome executives spend the entire day on the phone, desperately
working against time, trying to convince bankers, previously so solicitous and accommodating, that another
loan now is *vital* to the company's survival. The bankers, however, are icy. Exhausted, Jack Gallagher goes

to bed, leaving the other executives to make some final, fateful telephone calls.

March 31, 1982: At 1 a.m. Kenny calls Mike Phelps, assistant to Energy Minister Marc Lalonde. Phelps, in turn, wakes his boss with the grim news. "The message was that they were technically in receivership," Phelps recalls later. "The minister and Mickey Cohen [Deputy Minister of Energy, Mines and Resources] and I met early the next morning to discuss the issue in a more rational environment. Then the minister had discussions with the four banks. We certainly had an inkling that the squeeze was getting there, but the call came as a surprise to me."

April 5, 1982: Five days later, about to detail Dome's affairs to senior-level government officials in a conference room at Place Bell Canada in Ottawa, Richards says, "There is anything we would rather be doing than this." They then reveal their cash-flow projections, reciting from what the Dome group call their Armageddon file. The mandarins' reaction is described by one of the Dome team as "genuine horror". The prospect of a Dome collapse is so chilling to the federal government that it is deliberated on by Prime Minister Trudeau, Finance Minister Allan MacEachen, and Energy Minister Marc Lalonde. The political choice is difficult: the government has to weigh the unpopular course of bailing out a company that, in the public view at least, has already received too many favors from Ottawa and is teetering towards bankruptcy because of its own avarice, against the certain damage that Dome's collapse will do to the Canadian economy and confidence in Canada's banking system. As a highly placed banker said months afterwards, "It would have been a catastrophe if Dome had gone under. It would have been devastating for not only the government and the four banks, but also for the entire financial community in Canada." Officials in Ottawa later insist that the federal government became involved in Dome's affairs reluctantly, and only because the chartered banks were unwilling to carry the company without Ottawa's help. Relations between the overextended banks and the government are often testy.

Mid-April 1982: In a tense financial cliff-hanger, Dome is able to squeak out of its immediate problems at the eleventh hour by arranging to sell half of its Arctic fleet to its subsidiary, Dome Canada, for $200 million. Dome Canada, by then scraping the barrel itself for funds, makes an advance payment of $100 million to Dome Petroleum so that the parent can pay off its immediate obligations. The plan to sell half the fleet does not become known until statutory filings that Dome is obliged to make with the U.S. Securities and Exchange Commission are released in Washington, D. C., and this leaves a sour taste in the mouths of Canada's investment community. A number of analysts express private concerns that Dome is "milking" the assets of its offspring. The sale of the ships subsequently falls through, however, because Dome cannot persuade the Citibank syndicate to remove what are called "negative pledges" —legal undertakings preventing the assets from being pledged as security to others. But Dome is unable to repay to Dome Canada the $100 million that it received as advance payment on the aborted deal, and that debt is to remain outstanding more than a year later, strengthening a feeling that the Dome Petroleum–Dome Canada relationship is unhealthily intimate.

Bill Richards, who is president of both Dome Petroleum and Dome Canada* when the $100 million loan is made, responds hotly to the "milking" allegation. If Dome Petroleum had gone under, he explains, Dome Canada, because of various guarantees it had given to nervous Japanese investors, would have been responsible for a loan of $400 million the Japanese made to finance drilling in the Beaufort. It would also have lost its main asset: the right to explore on Dome lands.

In January 1983, Richards is replaced as president of Dome Canada by Louis Lebel, a former vice-president of Chevron Standard, a move widely interpreted as putting increased distance between the two Domes.

Richards said later, "The directors of Dome Canada were faced with the choice of seeing Dome Pete go down the drain and then being stuck with a $400 million liability, which could very well have brought Dome Canada down. My view is that it is not a matter of misuse at all; it's a matter of the Dome Canada directors making a decision [to advance the money] which events have proved to be the right one [since it saved Dome Pete from bankruptcy]."

Spring, 1982: More than once, over the next few months, a rescue deal is about to be completed when one or other of the banks balks. These problems arise, in part, because each bank enjoys a different level of security, and those with better security (the Toronto-Dominion and the Bank of Montreal) are more willing, if necessary, to put Dome into receivership than the Bank of Commerce and the Royal, which have more to lose. At Dome, Richards and his colleagues joke that the banks have a system which Richards calls "bastard of the week". "You have four banks and they rotate that position of 'bastard of the week'. Now sometimes one bank would be the difficult guy, then he will be pacified and the next guy will take over the role. It's entertaining in a way, you know. First of all we go around and visit this bank and they tear a strip off the other banks. All the other banks are a bunch of low buggers, you know, and they [the ones talking] are pretty good fellows. And you talk to the banks about the government and the government are just a collection of rogues. You talk to the government, of course, and they have the same view of the banks. And so the result is there is a tremendous amount of tension and conflict amongst them; I think there is a hearty dislike between the government and the banks generally."

Richards said he had some sympathy for the banks, because everything had gone to hell for them. "A number of companies are in difficulties; every aspect of our economy is in trouble. The banks must be under tremendous pressure when you consider that they are financing the whole of Canadian industry, and the whole of Canadian industry is going to hell: mining, forest products, construction, oil and gas. The banks are under

great pressure from a variety of different sources, and
the result is that, I guess, sometimes they succumb to
a degree of hysteria." With breathtaking casualness,
Richards adds: "It is obvious, there is no question about
it, the banks are very excitable."

The leaking ship is almost swamped; during the
spring, several Canadian bankers fly to Frankfurt on
a Friday night to pacify a worried German lender who
is threatening to call his loan, which would precipitate
an immediate collapse.

Mid-May 1982: Commerce Chairman Russell Harrison
goes to Ottawa on behalf of the CIBC and the three other
banks to meet Lalonde. They agree to a key federal
condition: that the government and the banks will be
equal partners in any arrangement to bail out Dome.
Ottawa also insists that the outcome must not only help
the banks, but return Dome to solid ground.

Late May 1982: A secret audit of Dome's operation,
commissioned by Ottawa and prepared by the Toronto
bankruptcy specialists Peat Marwick Mitchell and
Company, is handed to the government, spelling out
the major problems and possible solutions.

June 1982: Having considered the Peat Marwick report, the cabinet's priorities and planning committee
gives Lalonde and the Department of Energy, Mines
and Resources a mandate to negotiate a deal with the
four banks to assist Dome. The government, facing a
$20 billion-plus deficit, is adamant that its contribution
to any bail-out will not come from the government's
general revenues.

There are three main players in the drama that is
to unfold over the next three months. Marshall (Mickey)
Cohen, the Deputy Minister of EMR, who will shortly
move with Lalonde into the finance ministry, is in his
mid-forties, talkative, affable, and a tough negotiator.
Cohen practiced law in Toronto for ten years before
going to Ottawa as a special tax adviser in 1970. He
played an important role in energy negotiations between Ottawa and Alberta in 1974 and was the senior
bureaucrat behind the National Energy Program. The

second player on the government team is Edmund Clark, an economist in his mid-thirties, who was also one of the NEP's architects. He is known as an economic interventionist, and his doctoral thesis on "Public Investment and Socialist Development in Tanzania" has tainted him for all time as a "Red" in the eyes of the oil patch. The main spokesman and negotiator for the banks is Russell Harrison, who has been with the Commerce for thirty-seven years. He survived not only the Second World War, serving with the First Canadian Parachute Battalion in Europe, but also a number of bloody political battles in the boardroom as he rose to the top job in the giant CIBC.

June 15, 1982: Cohen flies from Ottawa to Toronto to begin discussions in earnest at Commerce Court. By this time the once-proud Dome has become so enfeebled, its bargaining position so weak, that the company's fate is debated between government and the banks, and Dome isn't even allowed in the room.

Late June 1982: Dome's affairs are now urgent enough that the Crown corporation PetroCanada is ordered by the government to give Dome's subsidiary Dome Canada a $100 million loan guarantee. Actually, the money is owed by the federal government to Dome Canada, because Ottawa has been tardy in paying exploration grants due under the National Energy Program.

Squabbling among the Canadian banks becomes intense. For a number of weeks the Toronto-Dominion holds up a deal because of difficulties with a $250 million loan guarantee it has obtained from Dome Mines, Dome Pete's major shareholder, to whom the oil company turned in January when it needed somebody to hold its hand. For weeks the TD will not even discuss the problem with the other bankers.

June 30, 1982: Dome's debt now stands at $7.03 billion. This includes $2.41 billion in long-term debts due within one year; $284.4 million in bank loans; $1.04 billion in accounts payable; the $100 million demand loan from Dome Canada; and $3.2 billion in long-term debt of various maturities. In addition Dome must raise

$250 million to meet its budgets for improvements and continuing operations.

July 7, 1982: The Commerce's exposure to Dome (and that other corporate invalid, Massey-Ferguson) is so great that rumors have begun circulating that the bank itself will be in jeopardy should Dome falter. Tongue-wagging comes to a head when Gary Lauk, a New Democratic Party member of the British Columbia legislature, and a former provincial economic development minister, makes a blunder of appalling proportions. Lauk, a lawyer and a bantam cock of a debater, tells the House in Victoria he has information that the Commerce will be in virtual, if not actual, receivership by October.

The effect of foretelling a bank collapse, like shouting "Fire" in a crowded theater, is electrifying and predictable. All night the Commerce's Vancouver regional office is besieged by calls from depositors seeking reassurance that their savings are safe. (In the midst of high drama, there is a moment of farce: CBC radio, in a broadcast from Toronto, attributes the MLA's remarks to Len Lauk, Gary's brother, who happens to be the Western regional head of the CBC.)

July 8, 1982: The alarm created by Lauk's speech continues throughout the day, and other financial institutions, even firms of chartered accountants, are pressed for advice. One 84-year-old woman, all her savings in the Commerce, faints when Lauk's speech is reported on the radio, and a bank vice-president, hearing of the incident, calls to reassure her personally that her money is safe. A number of elderly depositors walk out of their bank branches with up to $20,000 in cash. Competing banks and credit unions, normally solicitous for new business, rally to the Commerce's aid and try to dissuade people from making panic transfers from the Commerce. They too are nervous about the collapse of a major financial institution through investor panic. An official of the Royal Bank in Vancouver spends most of the day on the phone telling branch managers to assure anyone who comes in to switch accounts that there is not the remotest danger of their competitor going under. Still, in Victoria and in Kelowna in the Okanagan Valley of British Columbia, where there are large numbers of retired people, there is a "considerable" transfer of funds into the Royal.

The value of Commerce stock falls to a ten-year low on the Toronto Stock Exchange, and other bank shares plummet too. Harrison, along with other senior statesmen from the business world, is on Parliament Hill in Ottawa, counselling Prime Minister Trudeau on the government's six/five income restraint program, when an alarmed assistant passes him the urgent word that there is a run on the bank. No phrase in a banker's lexicon conveys greater terror, for the system depends for its survival on the confidence of investors. A bank *never* has enough cash on hand to repay all its depositors if they try to withdraw their money simultaneously; the entire fabric of the banking industry, and the welfare of the companies and even governments it supports, is constructed on the fragile but fundamental premise that Aunt Agatha *believes* that her $2,000 will be in the teller's drawer any time she asks for it—and so she doesn't.

Angrily, Harrison asserts that Lauk's statements are completely without foundation, and reflect a total lack of understanding of the Canadian banking system in general and the position of the Commerce in particular. "They represent a highly irresponsible act by an elected official," he thunders. Hours later, Lauk, badly shaken, withdraws his statement and apologizes.

Bank analysts are convinced that the Commerce and the other Canadian banks were never in any real danger as a result of their exposure to Dome, although the jeopardy from the panic among their depositors was real indeed. The Canadian banks had good collateral for their loans to Dome, but how much they would have got from a fire sale of Dome's assets in the event of a bankruptcy is a moot point, and undoubtedly there would have been significant losses. The bigger losers would have been some of the foreign banks whose loans were not as well secured.

Summer, 1982: Desperately searching for ways to raise money, Gallagher writes to Alberta Premier Peter

Lougheed asking if the provincial government is interested, perhaps through the Heritage Savings Trust Fund, in buying Dome's big block of shares in TransCanada PipeLines Ltd. Lougheed, facing an election in the fall, probably views such an offer to help Dome as a political liability. In any event, he does not bite. Later, the federal government proposes a plan to take over Dome's TCPL interest as security for its planned loan to the company, but Lougheed freezes at the idea of allowing Ottawa to grab control of the only transportation artery carrying Alberta natural gas to the big Ontario market. A group of Alberta gas producers, similarly alarmed that TCPL may fall into federal hands, makes tentative moves to purchase Dome's interest, but back off when they realize the price will be too high. In any event, Lougheed's stern rebuttal has discouraged any move by Ottawa to take over TCPL.

August 30, 1982: Dome announces that it lost $63.2 million, or 28 cents a share, in the first six months of 1982—its worst loss ever. Significantly, these figures do not include another $132.5 million in capitalized interest—money borrowed to pay interest on earlier debts. (This practice is not quite as bizarre as it might seem, and is used often in the financing of major projects when a financial return is not expected for several years.)

September 3, 1982: Dome's plight is causing the money markets to become jittery. The Canadian dollar falls more than four-tenths of a cent against the U.S. dollar, which analysts say is partly because of the bad news of Dome's financial difficulties.

September 7, 1982: Prime Minister Trudeau causes panic among Dome shareholders by saying in a radio interview that the government won't bail out Dome. "The government did not force the banks to lend money to Dome. The banks thought they were going to make a buck and Dome was investing this money because it thought it would make a buck. Are we going to bail it out? The answer is no, we're not going to bail it out." Dome's shares drop 45 cents—about ten per cent of their value—following the PM's statement. They rally in the afternoon following "clarification" of Trudeau's remarks by Lalonde, Dome, and the Bank of Commerce. Lalonde says Trudeau didn't mean to rule out all help, but merely aid in the form of a huge infusion of cash.

September 17, 1982: Dome's stock rises 25 cents to $5.25 on volume of more than half a million shares as rumors spread that Dome has successfully completed its refinancing. A Dome spokesman says the rumor is untrue.

September 22, 1982: Jean Chrétien, who has taken over the energy portfolio from Marc Lalonde, agrees to the principle of a bail-out arrangement hammered out over many months by bureaucrats and the Big Four. The ominous call goes out from Commerce Court in downtown Toronto to Calgary: "We have a deal for you."

By the time the phone call is received at the black glass Dome Tower on Calgary's Seventh Avenue S. W., Dome's fuse has burned alarmingly short: repayment of $1.3 billion is just a week away and Dome has no possible means of meeting that enormous bill.

Word is sent to the company's flight operations center at Calgary International Airport to have the twelve-passenger Gulfstream II executive jet prepared for a quick flight to Toronto, with an intermediate stop at Winnipeg to pick up other Dome directors. The date of the return journey is as uncertain as the company's future, and the two pilots and the flight attendant will have to stand by in Toronto for instructions.

Gallagher, Richards, group vice-president John Beddome, and finance vice-president George Watson, together with Calgary lawyer Maclean Jones, a Dome director, are soon flying east in the twin-engined jet. On arrival in Toronto they check into a series of adjoining suites on the sixth floor of the stately King Edward Hotel. Two elderly founders of the company, Bill Morton, now 76, and John Loeb, 79, fly in from the United States. Other directors converge on Toronto from Quebec and Ottawa.

Gallagher has dinner in the Bank of Commerce's elegant private dining room with CIBC chairman Rus-

sell Harrison, Rowland Frazee, chairman of the Royal
Bank, and Deputy Minister of Energy, Mines and Resources Mickey Cohen. It is during this tense meal that
the Dome chairman is told of the deal that the government and the banks have arranged for his company:
the immediate requirement to pay $1.3 billion will be
deferred and the government and the banks will make
available a further $1 billion to Dome to help the com-
pany straighten out its affairs. Gallagher merely toys
with his food as he listens apprehensively for the other
half of the deal. In return Dome will be required to
issue debentures which can be converted into common
shares of the company for $2 at any time during the
next decade. Gallagher is horrified: these are brutal
terms indeed, for Dome was trading at $5.25 only that
afternoon. He tells the bankers they will lose credibility
both domestically and internationally by supporting the
proposal. "I think what you will lose is immeasurable,
and [the action] will hurt you," he says. When he is
told that the price is the result of two appraisals that
have been made of the company's assets, he comments,
with some asperity, "Well, anybody can say 'zero for
this [asset] and zero for that' in a down market." Harrison retorts angrily that he is not living in the real
world: "Dome is bankrupt!" Cohen, although less abrasive, also feels Gallagher's view of Dome's value at that
time is quite unrealistic. But Jack Gallagher believes
it is obvious from the price that was set that some government bureaucrats are anxious to increase Ottawa's
stake in the oil industry by taking control of Dome. "I
don't think [Energy Minister] Jean Chrétien and the
politicians in general are as interested in that as some
of the stronger members of the civil service. I think
that's where the designs originated. I wouldn't name
who it could be, but it is certainly at the higher echelon
in the civil service, where there is the desire to increase
the national ownership, as exemplified by Petro-
Canada."

September 23, 1982: At a morning meeting in the
boardroom of Dome Mines the rescue proposal is presented to the full Dome board, and then Harrison makes
a surprise—and hurtful—request of Gallagher: the

Commerce chairman suggests that both Gallagher and Richards should be excluded from direct negotiations, to be conducted by only the "outside" directors of Dome— those who are not full-time working officers of the company. Harrison wants to deal with representatives of the shareholders and not management. Reluctantly, Gallagher agrees to this arrangement, feeling that perhaps there is some small bargaining advantage if the actual negotiators can insist that they must caucus with the entire Dome board before making any decisions.

September 23-29, 1982: Negotiations continue for almost a week but Dome has little firepower. Bill Mulholland tells Gallagher bluntly in private that the Bank of Montreal is prepared to put Dome into receivership the very morning after the default deadline if the company doesn't come to terms. His bank, he says, has already arranged for a team of lawyers to meet in Calgary at 8:30 a.m. on Friday, October 1, with all of the documents drafted to put a receiver in and take control of those Dome assets on which the bank has liens. "It's pretty difficult to negotiate with a gun at your head," Gallagher says soberly afterwards.

Despite its weak negotiating position, Dome's "outside" directors do manage to squeeze out improved terms: instead of the flat $2 conversion rate, conversion will be on a sliding scale from $2.50 to $5 over several years, and Dome's existing shareholders will have the right to buy $500 million worth of convertible debentures. Both the government and the banks will name directors to sit on Dome's board, and its capital spending and the sale of assets will be rigorously controlled by the company's new masters. If Dome does not spend enough on exploration in the Arctic, the government will grab its acreage there.

Bill Richards, who is acutely aware of Dome's vulnerability, congratulates one of Dome's directors: "You negotiated standing on a trapdoor, a hangman's noose around your necks."

Ottawa, facing severe financial problems of its own, arranges to borrow its $500 million contribution to the refinancing package from the banks, to be repaid from the proceeds of the Canadian Ownership Account, a levy that consumers are charged every time they fill up their automobile gas tanks or pay their natural gas bills. This tax, bringing in about $80 million a month, is still paying for PetroCanada's purchase of a string of gasoline stations in eastern Canada bought from the Belgian company Petrofina. When Petrofina is paid off, the levy can be used for the Dome debt. It will cost every man, woman, and child in Canada about $20 to pay Ottawa's share of the deal.

September 29, 1982: The so-called "bail-out-deal"—in fact, an agreement in principle—is signed at 8 p.m. in the 56th-floor boardroom of the Canadian Imperial Bank of Commerce at Commerce Court. There is a long-arranged big black-tie bankers' dinner this evening. Russ Harrison is already wearing his tuxedo; the other bank chairmen rush off to change once the agreement is signed.

September 30, 1982: Headlines give the impression that the banks and the government have already invaded the Dome Tower. In fact, the memorandum of agreement cannot come into effect for many months, until a detailed arrangement can be voted upon by the company's shareholders. The agreement is a curious document and contains so many gaping holes that a Sherman tank could be driven through it without difficulty. For example, it is contingent on Dome's foreign bankers putting more money into Dome and, in effect, assuming a significant part of the risk to which the Big Four Canadian banks have become so unwisely exposed. This does not sit well with the astounded foreign lenders, who have been kept in the dark and have only read about the so-called rescue and their own supposed participation in the newspapers. More than one describes the Canadian banks' attitudes as "imperious".

One foreign banker, with some exasperation, says his company will be prepared to assist Dome, but is not prepared to bail out the Canadian banks. "They are the clowns who got us into the mess that we are in by having those bottomless pockets a year ago. We will work on an assistance package for Dome, but the chartered banks have got to be in it. The proposal that is on the table is simply to take $500 million off their books and put it into our books, and that doesn't seem to help the company nor keep the chartered banks involved."

October 1, 1982: George Bennett, president of Boston's State Street Bank and Trust, sharply criticizes the deal. State Street and its clients have invested heavily in Dome. The firm owns 4.8 million common shares; its clients hold tens of millions more. "This is close to a conspiracy between the government and the banks to take advantage of a private company for their own self-interest in a situation they helped create themselves," Bennett says in an interview with the *Toronto Star*.

Within days of the deal, Citibank sends a telex message to the Canadian banks on behalf of its syndicate, rejecting any renegotiation of its own $1.8 billion deal with Dome.

Early October 1982: Predictably Dome shareholders are enraged and dismayed by the terms of the deal, and the word "rape" is commonly used to describe it. The magazine *Alberta Report* headlines its cover story "The Dome Rip Off." Harrison doesn't help matters when he suggests with breathtaking tactlessness that the shareholders should go down on their knees and thank God for the action of the banks and the government in saving their company. They are not inclined to follow his advice. In an interview with the *Toronto Star*, the Commerce chairman says, "Shareholders had absolutely nothing. They shouldn't be suing me [which one broker had threatened], they should be sending me flowers. If we'd liquidated Dome, then it would have been a disaster for common shareholders. As for 'poor old Dome' being under pressure, 'poor old Dome' held a gun to our heads. There's never been a company that's done as much harm to our country and its credit rating internationally as this company."

Referring to the role of the U.S. banks, he claims that they were "the least disciplined and the worst in the world. A pile of loans out of all this mess were unsecured. I don't think we were too bright in getting in so deeply, but they were downright stupid." Dome,

he says, was adept at playing one bank against another
to increase its lines of credit. In addition, the competition for the company's business got out of hand. "There
was a queue. It was a disease; everybody thought it was
Christmas." Harrison believes that the Dome situation
has taught the country's bankers an important lesson,
and that they will trim their sails in future, and he
announces that the maximum the Commerce will now
lend any given corporation—without more thorough
debate and study than has been conducted in the past—
will be $500 million. The major American banks, for
their part, are contemptuous of the Canadian Big Four.
One highly placed Canadian banker relates that the
New York banks were shocked at the way in which the
Canadians had loaned such enormous sums to one company. "They told me in no uncertain terms that if the
Canadian banks went to Europe [for funds], they would
not get quite the same reception they would have received the year before."

Late October 1982: Within a month of the Toronto
confrontation, however, the future begins to brighten
slightly for Dome. Interest rates have eased, money
begins flowing as oil and gas sales improve, and the
rigorous austerity measures put in place months earlier
begin to show their effect, with the result that the company can now face its future with greater equanimity
than it has for about ten months. Most importantly, the
"fuse" on its massive debt has again been lengthened,
albeit temporarily and to an uncertain length.

October 21, 1982: Jack Gallagher, sixty-six years old
and tired, but still resilient, refuses to be downcast and
searches for ways to keep the banks and the government at bay. Sitting stiffly upright in shirt-sleeves in
a large wine-colored swivel chair in his Calgary office
just three weeks after the traumatic encounter in Toronto, he discusses his feelings as he eats lunch at a
round marble-topped table.

"It's a great shock. I'm normally a very optimistic
person or I wouldn't be in this business," he explains
between mouthfuls of bran muffin, salmon, carrot sticks,

nuts, and asparagus spears. "I remember the head of Dome Mines, Jim McCrea, had had a disappointment in two mines the company had put a quarter of a million dollars into. One was in the uranium area of Saskatchewan and the other was in the Maritimes. They weren't massive amounts of money but he ended up with a heart attack and it killed him. I remember telling him, when he was so worried, even before the heart attack, 'You would die a thousand deaths in the oil business, because we drill a lot of dry holes and that's part of the game'. A lot of people don't have the optimism and therefore they shouldn't be in this business.

"Now I'm normally a fairly happy person. You ask my reaction [to Dome's financial collapse] after thirty-two years of building the company from scratch: isn't it a hell of a letdown? It is, but I don't let myself think in that direction. I feel that we'll work our way out of this, and so with that in mind I have no problem in carrying on. It doesn't hurt me. I always told my three boys, the most unimportant thing in life is making money. You can always get along with one suit if need be, but the important thing is to do something worthwhile, because that's what you will feel inside when your life winds down. 'What did I do that was worthwhile?' Not: 'How much money did I make?' You make money if you are successful, but that should never be the primary purpose.

"But I feel really sorry for the people in Dome who have had enough faith in me and the team to put all their savings in Dome stock. Approximately eleven to twelve per cent of the stock is held by employees, and therefore these people are affected far more by this particular move than would happen in an average company, where all it [the avoidance of bankruptcy] would mean is that their employment would go on. The government and the banks, I feel, don't recognize that one of the great strengths of Dome has been the high morale. It has attracted people with a lot of initiative. People who weren't interested in security, as such, but interested in doing things." Since Dome has signed a deal, or rather an agreement in principle, with the government and the banks, it is in a very sensitive position, and cannot make any overt move to wriggle out from that undertaking, no matter how distasteful, because the banks can demand their money at any time. Still, the company does nothing to discourage an organization called the Dome Shareholders Protection Committee, formed to fight the refinancing proposal.

Gallagher says he is receiving voluminous correspondence from shareholders, telling him of the support that is available for the company. "The shareholders want us to know that a rights offering to shareholders would be very acceptable, and they suggest that we can raise all the money we need in that direction. Our reply to them is that while we still have this overhang of $1.3 billion over our heads, it would be impossible to write a prospectus to cover a shareholder rights offering in which we would definitely have to disclose that at any time the banks could put us into default. We are hopeful as interest rates go down, as our cash flow increases, that we'll be in a position to possibly ask the banks, because of the shareholder reaction, to give us a conditional terming of those bank loans back to what they were before, conditional on a successful rights offering."

Gallagher's voice is strong and determined. As one of the worst years of his life draws to a close, it is very clear that Jack Gallagher, the entrepreneurial dreamer, although badly bloodied, has not given up the fight.

Chapter Fifteen

INQUEST:
WHAT WENT WRONG?

*"What is required is a 180-degree change in approach—
the virtues of yesterday are the vices of today.
Management must develop and apply the pedestrian
qualities of caution and restraint."*
—Bill Richards, November 1982

After the event, monumental blunders often seem simple to understand, and one wonders why those involved could not have foreseen the appalling consequences of their actions with the intelligence and clarity enjoyed by the rest of us. Trying to analyze Dome's pursuit of the Hudson's Bay Oil and Gas Company with the benefit of perfect hindsight strains credibility: how could a company so deeply in debt, so bloated by previous unregurgitated feasting, so committed already to a dozen grandiose projects with price tags in the billions, ever take on that monumental additional burden? But Dome almost got away with it, and it is worth recalling the cheering crowds that gasped in awe as it swaggered off in full battle cry. What went wrong? It was a combination of poor timing, bad judgement, politics, a rapidly collapsing economy, high interest rates, a conviction of the company's invincibility, and appalling bad luck that rained down with the sudden intensity of a tropical thunderstorm.

The simplest thing to understand is the motivation of Dome's architects, which can be explained in a single word: *greed*. The risk Dome took was not ennobled by any shining vision—a desire to develop Canada's fron-

tiers or a wish to secure the country's oil self-sufficiency. (The takeover didn't even increase the level of
Canadian *ownership* in HBOG, although it did change
the seat of control, since, ironically, the company directed from Stamford, Connecticut, already had a greater
percentage of Canadian shareholders than Dome.) The
dice were thrown and Dome's future was gambled simply to enhance the wealth and power of the aggressor.
There was no Holy Grail in prospect; just the spoils of
war to be grabbed by the victor. This is not a condem-
nation of free-enterprise goals which drive restless men
to promote their own enlightened self-interest doggedly,
for the incidental benefit of many; the question at issue
is how "enlightened" were Dome's actions and how well
were they executed?

In a macroeconomic sense, takeovers create no
wealth, no overall and redeeming benefit to society;
they merely shuffle the ownership of assets. The exceptions are when assets are badly managed or when
a merger creates economies of scale. Such cases are
rare, however, and were never argued as justification
by Dome. Indeed, the question of improved management would have been difficult to argue: HBOG was
regarded in the Canadian oil industry as a well-run, if
excessively conservative, company, with a good record
of profitability and exploration success, and Dome tried
hard, but in vain, to tempt its top executives to remain
after the two companies were finally merged. As for
economies of scale, many companies have discovered to
their discomfort that increased girth brings little more
than constipation. Bigger companies, like any large organization—government, the church, or the military—
require large bureaucracies to shuffle the inevitable
paperwork (or manage data systems), and the rapidity
of decision-making that previously characterized their
success becomes hamstrung by interminable memo-
randa and the scrutiny of myriad nitpicking commit-
tees.

Some Dome-watchers think the company had developed a dangerous belief in its own invincibility by the
time it took over HBOG. A competitor who views Dome's
fall with sadness says he saw an attitude creep into the
company that there was nothing they could not do.

"Some of their plans are so incredibly complex and technically so advanced...and frankly, they have done a lot. The men I know at Dome are very, very good. But it's almost as if people in the company were out-hustling each other. They were like people running downhill: they had their liquid natural gas export project, a new pipeline, the purchase of Davie Shipyards, all happening at once. There was a stage when there wasn't anything they wouldn't tackle. They had incredible self-confidence. Now that's been shaken. It seemed like back in the old days they were doing things all the time but they were doing them steadily and carefully and sensibly, and then these darned things kept getting better and bigger and bigger and there was increased velocity until it flew apart. I think there were so many individual things that had to be done that the wisdom at the top, Gallagher and the rest of them, didn't have enough time to look at everything carefully."

Actually, when the takeover was proposed to Dome's policy committee, comprising Gallagher, Richards, and the company's senior vice-presidents, Gallagher argued against it, standing alone in his dissent, although, significantly, he didn't say it was a bad deal. Then aged sixty-five, Dome's chairman was already feeling out of tune with the direction the company had been taking. He had suggested to his board of directors two years previously that he should move aside because the company was approaching a size that no longer appealed to him. However, the directors had all asked him to stay on, and he needed little persuasion because of his infatuation with the company's activities in the Beaufort Sea, and his desire, amounting almost to an obsession, to see oil produced commercially from the Arctic. But as he neared retirement age, he took a less active role and had shown a willingness to defer to others.

Gallagher argued against the takeover for several reasons—because of his dislike of gigantism, and because he felt Dome still had some cleaning up to do as the result of three earlier oil company acquisitions, Siebens, Mesa, and Kaiser. Also, he had made a telling point in an address to the Toronto Board of Trade in January 1981, just a few months before Dome began its pursuit of HBOG. Discussing the National Energy

Program, he said, "Few Canadians will quarrel with the objective of having a higher degree of Canadian ownership in the industry. However, I suggest that rather than buy out multinational companies with developed oil and gas reserves in Canada that already come under the all-encompassing regulatory and taxation controls of the federal government, it would be preferable to use those funds to find and develop more hydrocarbon reserves in Canada." If his own company had followed that advice it would have been spared much misery. HBOG *did* have developed oil and gas reserves in Canada and was *already* well regulated and taxed. Dome spent about $4 billion bringing it under its own wing but in so doing added not one drop of oil to Canada's reserves; in fact, the financial difficulties that it incurred retarded the exploration activities of both Dome and HBOG.

Despite Gallagher's stated reservations, which he labels emotional rather than logical, he went along with the deal because he did not want to dictate the type of company the younger men would be running after his departure. "Certainly I wasn't in favor, but I said I wouldn't oppose it if it could be done on a friendly basis."

Gallagher later accepted responsibility and stressed that he was in a position to prevail but did not do so. "I was still chief executive and I allowed it to go ahead, and therefore I am responsible for it because you can't say, 'Well, I voted against it and still let it go ahead.'" A Calgary oilman who knows both Richards and Gallagher well says, "I am sure Jack could have called the shots. He is a pretty determined guy. But arguing with Bill Richards is like trying to swim up Niagara Falls. I suppose somebody could calculate that it is mathematically possible. But Richards has such a quick mind and is such a tenacious arguer and is so confident that I suspect although Jack didn't want to do this he was overwhelmed by Bill's persuasiveness and machine-gun approach. He was probably worn out to the point where the guy said, 'Damn it, I don't want to do it but if the rest of you guys want to do it, okay.' There would at least be a tacit approval. I think Jack was just overwhelmed."

Dome's decision to acquire HBOG was taken in a climate of great inflationary expectations. Hundreds of companies, and many thousands of individuals, subscribed in 1981 to the prevailing view that since the purchasing power of money was being eroded rapidly, the really *smart* initiative was to buy tangible assets—gold, silver, oil paintings, condominiums—which were increasing sharply in value. Buy today, borrow if you need to, and tomorrow will take care of itself, was the conventional wisdom. Inflation, the bug that drives individuals to buy gold or antique cars as a hedge to preserve real wealth, is the same virus that spreads takeover fever among companies. Dome was not the only firm engulfed in the scourge that swept through Canadian industry in the late 1970s and early 1980s, although it was by far the most celebrated victim. Boardrooms across the country are hung with black crepe as the result of the takeover fever, when it seemed so easy (and cheap!) to use borrowed money to gobble up assets—real estate, timberlands, oil fields—for a fraction of their expected worth a few years hence.

Of all the prospects for raw materials which beckoned to investors during the boom, oil seemed to have the most allure, with price projections reaching up to $90 a barrel by the end of the 1980s. The banks were as bedazzled by the potential value of oil and gas in the ground as were the oil companies themselves. The decisions they took, involving mind-boggling sums, were not made by junior loan officers or lowly branch managers, but in wood-panelled offices by cigar-smokers in $900 suits: the cream of Canada's banking elite, who behaved with incredible recklessness. Describing how the banks scrapped among themselves for the biggest share of the booming Calgary oil business, Jim Gray, Executive Vice-President of Canadian Hunter Explorations Ltd., says, "The banks were holding breakfasts and lunches in their dining rooms. They were pushing credit on people and they admit it. When I grew up in Northern Ontario, you believed your doctor, you believed your minister, and you believed your banker. If a banker said, 'I will lend you $30 on the basis of your income from your newspaper route,' well, you really knew that it was safe to borrow $30 because who could be more conservative than your banker? In the late

1970s, bankers said, 'I will lend you $100 million.' Well, there was a kind of sympathetic understanding in that. You thought, 'Jeez, if my banker is going to lend me $100 million, then that's really a safe thing to borrow; he knows about interest rates and he knows about my equity and the collateral.' The banks moved some aggressive people out here and they have their own oil and gas departments. In some cases the banks were seeing more oil and gas reserves than we were as an industry. But I don't blame it all on the banks. They lost a little of their objectivity—we all did; it was heating up too quickly."

All would have been well for Dome if the economic boom had continued a little longer; but as it transpired, the crystal balls were badly out of focus. In 1980 and 1981, Ph.D. economists, working with sophisticated computer models to predict medium-term economic growth, accomplished the same degree of scientific exactitude as a bunch of giggly five-year-olds attempting to pin the tail on the donkey. Instead of birthday cake, their faces were covered with egg.

Until its celebrated financial difficulties Dome's attitude towards the control of cash flow was quite casual, and its executives never lost sleep trying to balance Dome's monthly income with bills coming due. This dangerous attitude arose out of the company's expansionist orientation. There was a strong emphasis on making good economic decisions (Dome has a large economics department staffed with talented, intelligent people) and the belief that if those decisions were made wisely, such mundane matters as a balanced cash flow would follow naturally. It never paid dividends to its shareholders or taxes to the government, choosing instead to plow back all its resources to strengthen the ever-growing company. With abounding self-confidence it believed that whatever difficulties arose, it could always arrange a quick financial "fix." The company became addicted to debt and was oblivious to the dangers of this addiction.

Undoubtedly one of the important factors that brought Dome to its knees was unprecedentedly high interest rates, which peaked in the late summer of 1981. But even earlier, while Dome's policy committee was still debating whether it should go after HBOG, the cost of money was anything but cheap. At the beginning of the year, the benchmark Royal Bank prime rate (offered only to the most credit-worthy corporate customers; Dome, already massively indebted, didn't quality for prime) was 18.25 per cent, and it remained at this level until March 23, when it dropped a modest half percentage point. When Dome launched its opening salvo against Conoco on May 6, the prime rate was 18.5 per cent, and two days later it jumped to 19.5 per cent, where it remained when Dome's deal with Conoco was signed on June 1. But these symptoms of problems to come were ignored until too late. In the weeks that followed, Dome executives read their morning papers with increasing apprehension. They had strapped the company to a torture rack, and four times within the next ten weeks, interest rates were forced up another excruciating notch, until they reached a life-sapping 22.75 per cent on August 7. At those rates Dome was paying well over $1 million *a day* simply to service its *new* debt.

Ironically, Dome had thought the acquisition of HBOG would assist Dome's cash flow. When, on June 26, 1981, five Dome directors—Gallagher, Richards, Maclean Jones and Fraser Fell, both lawyers, and Ottawa energy consultant Marshall Crowe, a former chairman of the National Energy Board—joined the board of the Hudson's Bay Oil and Gas Company, HBOG officers presented a ten-year financial plan which gave Dome a dangerously cozy feeling of security. As it turned out, HBOG's expenditures over the coming year were significantly understated, and revenues were unrealistically optimistic. Richards said later of those financial projections, "They were dreadfully inaccurate, but we did not have anything else to look at. If you only have one piece of paper to look at, you attach too much importance to it because that's the only thing you have to judge. I don't want to attribute wrongful motives. They were misleading...."

Because HBOG's projections had always tended to be extremely conservative, Dome was inclined to believe them. "We had no choice, we had no way of checking them. It was not until January 1, 1982, when I became president of HBOG, that we could get into the guts of the thing," said Richards. Directors never receive financial information in great detail. "You have to get your people, who have command of the detail, right into the guts and bowels of the thing. You have to have a guy talking to the production engineer and saying, 'Christ, you are projecting fifty barrels a day from that well, how the hell do you think it is going to make that much?' You have to get into that nitty-gritty, and of course boards of directors don't so that."

One reason its new acquisition never became the "cash cow" Dome had expected was the purchase by HBOG of the Cyprus Anvil Mining Corporation, which was to demand dramatically unforeseen financial nourishment. At that first boardroom encounter on June 26, HBOG Chairman Gerry Maier, who had been promoted to chief executive just fourteen months before, told his new masters that HBOG was within days of making an offer to buy Cyprus, one of the world's largest producers of lead, silver, and zinc concentrate, which had a huge operation at Faro in the Yukon. Sitting around the HBOG's board table for the first time, the Dome directors were asked if, as HBOG's new controlling shareholders, they had any objection to the purchase. This was new territory indeed for Dome; time was short and they were being asked to make a snap decision involving half a billion dollars and about an industry they did not understand. Gallagher sought advice from Dome Mines of Toronto, and, according to Dome insiders, they answered in effect that Gallagher should not touch Cyprus, which had already begun to lose money, with a barge-pole at the price asked. Strangely, though, in light of this advice, Dome did acquiesce in the purchase, for which HBOG paid $345.3 million and assumed about $150 million of debt.

The decision was an incredible disaster, coinciding with a dramatic collapse of the international metals market. Just five months later, Cyprus had 1981 operating losses of more than $10 million, and for 1982 these were estimated at more than $25 million. In September 1982, Dome decided to shut down the mine until the next spring, all but paralyzing the Yukon, where it accounts for forty per cent of all economic activity.

About a year after the mining company was purchased, an internal evaluation by Dome in the depressed market of 1982 put a value of $130 million on Cyprus *before debt*. It was an asset Dome would have difficulty *giving* away. Almost casually, through the Cyprus purchase, Dome had acquired yet another half-billion dollars in debt on which it had to pay interest, and had become the owner of a company, alien to its own industry, that was hemorrhaging badly.

Asked why he did not veto the mining company purchase, Gallagher replied somewhat disingenuously that it was made by HBOG and not by Dome. Reminded that Dome had five directors on the HBOG board at that time, he said the deal was virtually complete when Dome was advised of it, and in any case Dome could not *control* the board. This was a literal statement that ignores the commanding influence which representatives of fifty-three per cent of the company's shares would undoubtedly exercise over the other directors if they should voice strong arguments against a major new capital acquisition. Gallagher insisted, "It was presented with all the economics showing how good it was, the studies that had been made on the markets. All the work had been done, it had to be closed within a day or two, and we did not feel we were in a position to challenge something they were going ahead with. And if we hadn't moved onto the board on that particular day, they would have gone ahead with it anyway. We had an opportunity to object, but we didn't have control of the board and we didn't do any of the work that went behind appraising it. In hindsight obviously we should have said, 'To hell with it, we haven't had a chance to look at it.' But the time frame was so short."

This may be an example where Dome, having so much on its plate, was simply too preoccupied to attend to minutiae like the trifling purchase of an unprofitable mine for another half-billion! A contributing reason, suggested by a former Dome executive, is that Gallagher at that time was deeply anxious to recruit HBOG Chairman Gerry Maier, an executive of undoubted operating talent, as a stablemate (or even a replacement, perhaps) for Bill Richards, whose entrepreneurial cavalry charges did not leave him time for the routine

direction of Dome's increasingly complex operations.
For this reason, possibly, Gallagher did not want to
upset Maier by interfering with the Cyprus purchase.
As a result of this incautious move, Dome became entangled in the chronically rickety economic affairs of
the Yukon and yet another problem was added to the
burdens of the harried Dome management, already
working fourteen-hour days and weekends to clear a
muddle of escalating debts.

References to warfare keep recurring in discussing
the HBOG takeover, and in one sense in particular the
analogy is apt: the need to maintain momentum, the
vigorous onward thrust towards a well-identified target. Having won a substantial foothold in HBOG on June
1, Dome lost its drive and was beset by uncharacteristic
indecision as to its next move, and for about two months
it alternated almost daily between choices of action as
to how it should capture HBOG's cash flow. During those
crucial weeks in the summer and early fall of 1981,
unexpected setbacks of great severity hit the oil and
gas industry. It is now clear that Dome did not realize
just how fast its fortunes were declining in that period.
Bill Richards, reflecting on this later, said, "I guess the
thing that I have learned is that we should have tried
to be more responsive to the changing circumstances
as they occurred. Having said that, I don't know how
the hell you could do it when you are in the midst of
something like that, and your data is inevitably two or
three months old. The LPG [liquefied petroleum gas]
prices were all going to ratshit, but you really didn't
get a full sense of the scope of this until maybe a month
or two later; and the same with production cutbacks
and all those forces. I guess in a business sense, had
we been better set up to monitor those changing circumstances as they occurred, we might have been able
to react more quickly. That's the element [of the take-
over] where I think maybe we could have done better."

Dome's internal deliberations at this poing give a
clear insight into its potential ruthlessness. There were
two possible ways to gain access to HBOG's cash flow: it
could persuade the remaining shareholders of HBOG to
sell their interests to Dome, or it could tap HBOG's wealth
by inter-company deals, forcing HBOG to buy various

Dome assets for cash. This latter course was fraught with difficulties, however, since such deals would be scrutinized coldly by regulatory authorities, and Dome would be under pressure to demonstrate that its deals were fair business undertakings of commercial benefit to both parties. Dome's lawyers believed this plan could work, but warned that the company could fall into a legal morass. Dome decided it would be quicker, and less messy, to buy out the remaining shareholders.

In early August 1981, when Dome decided it would go after the remaining shares, it was necessary to play the coquette, interested but not eager, an unusual role for the impulsive and forceful company. There was good reason to play games. An indication of undue interest would drive up the price demanded by HBOG shareholders, and also, since it would be offering "Dome paper" in exchange for HBOG stock, Dome wanted to delay consummation until it received its 1981 Beaufort Sea drilling results, which it hoped (vainly, as it happened) would enhance its stock price and strengthen its bargaining position.

These events, and others, slowed Dome's momentum, and time, a precious but unrecognized element, began to run out. If Dome had completed the second half of the takeover as expeditiously as the first part, it would have been in a legal position to sell off portions of its acquisition profitably in the summer of 1981. Customers were nibbling and their prices sounded good. But eight months later, when Dome finally concluded the deal, the economic climate was vastly different, and potential buyers, who had seemed eager in the warmth of late summer, were decidedly skittish the following chilly spring. One insider estimated that if Dome had seized HBOG's cash flow three or four months sooner, it might have commandeered an extra $60 million or $70 million, but more importantly it would also have gained a quicker appreciation of the difficulties to which HBOG was exposed as a result of the unwise Cyprus purchase and the large financial commitments HBOG was making in Indonesia and Brazil. Dome had waited too long, and by the time it finally merged with HBOG on March 10, 1982, it was facing almost immediate bankruptcy.

In addition to its other problems, Dome also became

the victim in late 1981 and early 1982, along with other
oil companies, of political bloodletting between Alberta
and Ottawa. The announcement of the National Energy
Program on October 18, 1980, started a fractious dispute over energy pricing. While the battle was fought
over jurisdictional issues, Alberta protesting that the
federal government by imposing a wellhead tax on oil
and gas revenues was intruding into an area of exclu-
sive provincial jurisdiction, at heart the row concerned
an issue far less esoteric than constitutional subtleties:
at stake were many billions of dollars in taxes from the
oil industry that Ottawa saw as a salve for its own
revenue wounds and Alberta perceived as a raid on a
birthright belatedly received. It was, in Premier Peter
Lougheed's folksy phrase, "like having strangers take
over the living room." To discourage the visitors, he
planned moves to show them how unwelcome they were:
he halted negotiations on additional tar sands plants,
and ordered that between March and September 1981,
oil companies in Alberta must cut back production by
fifteen per cent in three stages each of sixty thousands
barrels a day. This tactic backfired: world supplies of
crude oil were plentiful at the time, and so refiners in
Eastern Canada were able to arrange additional imports to replace the loss from Alberta. Furthermore,
Ottawa subsidized their extra cost by an increased
charge at the gas pumps, which they cleverly called the
"Lougheed Levy". The unseemly dispute ended in early
September, but even when Lougheed and Prime Minister Trudeau clinked champagne glasses in Ottawa to
signify a truce, the oil industry's difficulties were far
from over. The new revenue-sharing formula had an
even harsher impact on the oil producers' revenues than
the NEP. And there was an added irony: during
Lougheed's cutbacks, refiners in Eastern Canada had
discovered that it was possible (with the benefits of
those subsidies at the gas pump) to import foreign oil
cheaper than that produced in Western Canada—naturally they signed a few long-term contracts! It is one
of the silliest episodes of recent Canadian history that
motorists were subsidizing foreign imports while Alberta oil was shut in the ground. Alberta producers
were obliged to prorate their production for many

months. The effect on Dome was serious: lost production of about 10,000 to 15,000 barrels daily during the summer of 1981 and a loss in 1982 equivalent to 25,000 barrels a day for about three months.

There are several actors in Dome's financial tragedy and they each share, in varying degree, blame for the disaster that befell the company. The federal government, while it had no direct responsibility for Dome's undoing, certainly created through the National Energy Program the political climate in which Canadian oil companies were encouraged to become raiders, eagerly snapping at the heels of the multinationals. Ottawa did not force Dome to become a predator, but it certainly opened the door to the chicken coop.

The banks, whose leaders are prone to solemn pontification on the nation's ills, behaved with wretched foolishness, out-hustling each other for Dome's business. Like bartenders confronting a known alcoholic, they gleefully offered him a succession of doubles and then, when he collapsed in a stupor, became prune-faced preachers of temperance.

Dome's management, who proposed the takeover, and its board of directors, who authorized it, carry the biggest blame: in their unrestrained desire for growth they were blind to the dangers to which they were exposing the company. They acted freely and recklessly; no one forced their hand when they embarked on a wild adventure without parallel in recent Canadian history, in the full expectation of glory, and fell ignominiously.

One of the saddest aspects of Dome's collapse is the effect it has had on the morale of its employees, who have suffered severely. Dome built a stable work force through very attractive stock options or share purchase schemes, and many long-time employees became very wealthy indeed. At one time, in fact, the company was reputed to have more millionaires on its staff than any other corporation in Canada, and one secretary retired, after years spent pounding a typewriter, with a nest egg of shares worth a quarter of a million dollars.

Employees were hard hit by the fall in share value (from $25 in early 1981 to less than $3 in October 1982), and many fortunes were practically wiped out over-

night. One executive nearing retirement became
$100,000 richer or poorer with every $1 fluctuation in
Dome's share price. Many employees, who had moved
to Calgary, attracted by the booming oil industry, and
bought homes at the height of the phenomenal real
estate boom about 1981, saw both their homes and their
stock values plummet simultaneously, and even, if they
managed to hang on to their jobs, suffered cuts in pay.

Dome employees speculated heavily in the exploration subsidiary, Dome Canada, with numerous people
mortgaging their homes, because they could see big
riches to be made. They were cruelly disappointed: having paid $10 for their shares early in 1981, they found
these to be worth about only $4 two years later. A for-
mer employee says there was tremendous peer pressure
to buy the new issue. This happened inadvertently, since
the last thing management wanted to see was its employees hurt by a speculative investment. But, because
it seemed the offer would be oversubscribed, the company set aside a number of shares for Dome employees,
who bought about $30 million worth. The Toronto-
Dominion Bank, which has a branch at the foot of the
Dome Tower, came into the offices and offered loans to
facilitate the stock purchases. Many employees, believing they were availing themselves of that famous
"chance of a lifetime", got caught up in a cheerleading
atmosphere and practically entered into a competition
to see who could buy the most of the new issue. Many
dispirited Dome employees now bitterly regret the enthusiasm of the moment.

Perhaps it is fitting, at the end of an inquest into the
HBOG debate, to let the architect of the deed offer his
perceptions. Bill Richards was in a philosophical but
unrepentant mood in the fall of 1982. He put the company's difficulties down to bad luck, not mismanage-
ment. "I fully believe in luck in this business. When
you are lucky, you are happy to be the beneficiary of
good luck, and you can go out with your chest out and
say, 'What a smart guy I am.' So when you are the
recipient of bad luck, it seems to me that you have to
say, 'Oh well, obviously I made a hell of a mistake.' The
concept was perfect. Classically and technically it was

a magnificent takeover every step of the way," he insists defiantly. "It worked out great...it was just that the whole world went all to hell in the meantime."

By November 1982, Dome had learned some bitter lessons, and Richards had unbuckled his six-shooter and changed from his gunslinger's costume into the uncomfortable charcoal grey of corporate rectitude. In a speech entitled "A Prescription for Canada's Economic Ills", he said, "Most of us have spent our entire business careers working in an economy which was expanding or at least relatively prosperous. It will be a true test of management to see how well we can respond to this radically changed business environment. What is required is a 180-degree change in approach—the virtues of yesterday are the vices of today. Management must develop and apply the pedestrian qualities of caution and restraint. The near-term solution must be to control tightly spending on both capital and operating costs, in order to develop a lean, efficient operation. The challenge to management is to provide a dynamic response to the drastic change which has occurred in the economy over a period of less than a year.... There never was a time when stable management with steady nerves was more required."

Richards was probably aware of a profound irony as he spoke those words. In its early days Dome did practice "the pedestrian qualities of caution and restraint." It took risks, but it also hedged its bets by investing most of its money in solid, unglamorous enterprises that ensured a firm foundation for the company. Then the adventures became more flamboyant and the underpinnings less secure. Now it had come full circle: its high-flying risk-taking days were over and it had to begin the slow and painful task of reconstructing The House That Jack Built.

Chapter Sixteen

DOME LOOKS AHEAD

Almost seven months after the "bail-out" crisis of September 1982, an era ended at Dome. On Friday, April 8, 1983, Jack Gallagher walked slowly from his splendid corner office on the thirty-third floor of the Dome Tower in Calgary and into a meeting room where he informed assembled executives that he was formally relinquishing his position as the company's chief executive officer—"dictator" in his own description.

He had thought about the event often; indeed he had wanted to leave office several years earlier but had been persuaded to stay. Still, to his surprise, when the time came he found the action unexpectedly "wrenching". It was a particularly poignant moment for the restless man who had devoted thirty-three years of intense effort to nurturing a one-man wildcat oil company into the third-largest private-sector corporation in Canada. Dome had grown far beyond his wildest hopes—or desires. It held the largest spread of oil and gas lands in Canada, with interests totalling more than eighty million acres, and in 1982, despite its much publicized difficulties, the company and its associates had drilled about sixteen per cent of the exploratory well footage in Canada in its never-ending search for hydrocarbon wealth.

An executive who returned to the Dome Tower the next day said it was a very strange feeling to be in the building and realize that after more than three decades Jack Gallagher was no longer in command. Gallagher had said some time before that he would step aside later in the year, but his departure from the top job was hurried to pacify the bankers who still held Dome's fate in their hands. They no longer wanted to talk to him, preferring to deal with the company's "outside" directors—those who were not members of the company's management structure. Gallagher remained as chairman of the board, but, divorced from the authority of chief executive, the chairman's position had become largely titular. Gallagher also stayed on in his spacious office as chairman and chief executive of Dome Canada, which conducts most of the exploratory drilling on Dome lands. Directing the search for oil was much more to his liking than haggling with hostile bankers. Gallagher maintained a brave front that nothing really had changed: he still put in full days at work and was in the office at weekends, he said. But others in the executive ranks believed it was only a matter of time before he relinquished his remaining responsibilities too; he was being eased out graciously as befitted a man who inspires an almost universal affection in Calgary oil circles. Gallagher was sixty-seven and tired, but although his voice sounded at times almost feeble, he still managed to rekindle his old spark of infectious enthusiasm when he talked about his future plans, including one of his pet projects—promoting reform of the Senate.

Gallagher left office well provided for. On his retirement as chief executive he began receiving a fee of $27,500 a month, to continue for eight years if he remained as a consultant to the board of directors. If the agreement is terminated before the end of eight years, he will get a lump-sum payment of $2.6 million. The company's public affairs department spent about a week justifying this arrangement to many people who felt that the deal was a bit rich for a company still facing bankruptcy. Gallagher, however, had not drawn a salary from Dome since 1967; by then he was comfortably well off and able to finance what he calls his "modest" needs from his previous savings and investments. The termination agreement, he explained, represented money owing to him in lieu of salary, plus compound interest, for sixteen years.

When Gallagher resigned, Dome's directors appointed a committee to act "in the capacity of chief executive" until a replacement could be found. This group was headed by Dome director Frederick W. Sellers of Winnipeg, the 52-year-old president of Dionian Industries Ltd. and chairman of the Canada Development Corporation, where Gallagher used to be a fellow director. Calgary buzzed with speculation as to who would succeed Gallagher, much of the tongue-wagging centering on John Beddome, who was promoted from group vice-president to the position of executive vice-president and chief operating officer, although insiders said he was certainly not lobbying for the job.

Gordon Harrison was not in the running at all. Earlier in 1983, Harrison, who as president of Canadian Marine Drilling had directed the company's northern operations for eight and a half years, left his $193,333-a-year job for the number two spot and an ownership slice in a privately owned oil company in Houston, Texas. He departed with mixed feelings, typically still tidying up loose ends of his Dome job until just a couple of hours before his plane left Calgary for Houston one Saturday in January. He admitted he had thought of leaving some months earlier but did not want to give the impression of deserting a sinking ship. By early 1983, however, he felt that the vessel was more buoyant. He insisted that he was not quitting because of any diminution of Dome's efforts in the Arctic: the company's Northern technology had evolved to the point where Dome could successfully extract and transport oil from the frontier once it was found in sufficient quantities, and he felt no pressing need to remain until the Arctic actually yielded hydrocarbons in commercial quantities. It was time for him to move on. On Harrison's departure, the frontier operations were centralized under the growing fiefdom of Beddome.

At that point the future of Bill Richards was very uncertain. His knowledge of the company was comprehensive and many insiders were hoping he would remain in the president's chair, although it seemed apparent that his chances of ever taking over the chief executive's job had evaporated in the heat of the HBOG misadventure. Gallagher, for one, did not consider the man he had picked out from the company's legal department and advanced to the presidency to be a suitable successor; he had already offered his own job to

Bill Daniel, president of Shell Canada Ltd., who had turned him down. Gallagher hinted at the time of his own resignation that Richards would be the next to go. In interviews, however, he punctiliously sidestepped criticism of Dome's president or the others in the senior management ranks who had supported the disastrous HBOG takeover which he had opposed.

By the early summer of 1983 Dome was still a very sickly patient with a long convalescence in prospect, although it had certainly improved from the previous fall. The dramatic decline in interest rates that began in 1982 and continued into 1983 was a remarkable tonic. Dome was again paying its bills and interest on its debt; it had sold off half of its holdings in Trans-Canada PipeLines Ltd. for $146.8 million and was apparently preparing to dispose of its remaining TCPL shares. Still, its position was precarious and it could be placed in bankruptcy at any moment.

Its relations with its bankers remained its prime preoccupation, and every morning promptly at 8.30 a.m. Finance Vice-President George Watson conferred with a special management committee on the latest developments. Completing the financial restructuring was like unscrambling an omelet. The banks had to be satisfied about the safety of their money; the shareholders—over 43,000 of them—dismayed by the threatened dilution of their investment, had to give their approval to any agreement; and the federal government would have to put a legislative seal on the Dome rescue package, if it was resorted to. Each player's objective was different and often seemed irreconcilable with those of the others. Dome's management wanted to retreat from the deal negotiated so painfully the previous fall and so keep the government and the banks out of their boardroom, using the arrangement merely as a safety net if all else failed. The banks—more than forty of them around the world—ideally wanted their money back, or, if they could not get it, they sought the very best security for their loans and did not flinch from body-checking their competitors to gain advantage. Those with the superior security—for example, the members of the Citibank consortium, whose loans were backed by oil and gas production—were disinclined to offer aid to fellow bankers whose flanks were imprudently exposed. The squabbles were endless, debilitating, and often intemperate, and for months produced no results. But the Big Four Canadian banks, who were parties to the "bail-out deal" and were in greater jeopardy, declined several opportunities to pull the plug on Dome and allowed each fresh deadline for repayment to pass.

However, even as Dome sailed along cautiously with the forbearance of most of its bank creditors, the company was still exposed to unexpected squalls. A dramatic upset occurred in the spring of 1983, and Dome's ability to navigate away from the financial shoals showed that it had not entirely lost its old gift for adroit maneuvering. The problem arose through its ownership of the big mining company Cyprus Anvil, which has a lead-zinc mine at Faro in the Yukon. Long before Dome's involvement with Cyprus, the mining company had borrowed $130 million from a banking group which included the Mitsui bank of Tokyo. Repayment to Mitsui of Cyprus's loan became due on May 13 and the Japanese demanded repayment. In the context of Dome's financial problems the amount was trifling, a mere $300,000, which the company could have repaid without difficulty. However, when it signed its agreement-in-principle with the Big Four Canadian banks and the federal government in September 1982, it promised not to make payments of principal to anyone. If it paid Mitsui, the September agreement would be in tatters; yet if it failed to make the payment demanded, Mitsui could declare Dome in default, which would trigger cross-default clauses in all of its banking agreements, with the renewed threat of imminent bankruptcy.

Dome nimbly outran the Japanese by selling part of its controlling interest in Cyprus to Dome Canada, its subsidiary, for the grand sum of $20, on the understanding that it could repurchase control for the same sum should it wish to do so in the future. The maneuver successfully isolated Dome from the financial problems of Cyprus, another corporate cripple: since Cyprus was no longer technically a direct subsidiary, Dome was not responsible for the payment of its debts. Mitsui could now make its demands only of Cyprus, and even if it put the mining operation into bankruptcy, that would not bring the entire Dome empire of cards tumbling down. Surprisingly, it was all quite legal, if a little sleazy; but Dome could not afford the luxury of a squeamish stomach.

With every asset worth selling nailed down as security for *someone's* loans, Dome's attempts to dispose of parts of its unwieldy empire were horrendously complicated. A prime asset, and candidate for sale, was Dome's 40 per cent shareholding in Dome Mines Ltd., an Ontario goldmining company which, in the early summer of 1983, had a market value of several hundred million. Dome Petroleum's relationship with Dome Mines is complicated. The mining company owns 25.9 per cent of the oil company and the oil company holds 40 per cent of the mining outfit. This corporate contortion (which means in effect that Dome owns 10.4 per cent of itself!) was arranged some years ago as a successful defensive fortification against the takeover of either company by outsiders. But the arrangement did not make the extrication of Dome Mines as a saleable asset any easier.

However, faced with difficulties that would have paralyzed most corporations, Dome continued to play the role of vigorous entrepreneur. Even as it battled for financial survival, it was successfully promoting a project of enormous proportions—to export liquefied natural gas from Alberta and British Columbia to Japan, through a new port to be built specially for its purposes near Prince Rupert on the B. C. coast. The scheme, which had first been unveiled in October 1980, calls for a new 550-mile pipeline, a natural gas liquefaction plant to cool the gas to $-260°$ F., at which temperature it occupies only a six-hundredth of its gaseous volume, and five enormous ships.

The British Columbia government, which had been lobbied relentlessly for many months, officially endorsed Dome's multi-billion-dollar gas scheme in July 1982, at which moment Dome would have had difficulty in financing, from its own pocket, the purchase of a Coleman camp stove! No matter; it was all being done with other people's money. If Dome had found a money tree in Ottawa, it also located a similar specie in Japan, for it has a tentative arrangement to borrow from the Japanese about $2.4 billion, to be repaid over ten years from revenues from the project.

The scheme has received initial government approval in Canada, and, despite some recent nervousness in Japan over Dome's financial strength and its ability to deliver on the deal, the project (for which Dome still needs to recruit one or more partners) appears to have a fair chance of success. Dome, clearly, is not about to roll over and die.

At the time of writing (the early summer of 1983), trying to see Dome's future was like attempting to penetrate the opaque Arctic fogs that so frequently blanket its Northern exploration operations. A few outlines were discernible, however, and other shapes could be guessed at. While much remained unclear about Dome's future, bankruptcy appeared to be a fast-receding possibility.

The important meeting at which Dome's shareholders would be asked to vote on the re-financing package was originally scheduled for the spring of 1983, but it was postponed while Dome's management worked desperately to offer shareholders a deal far more palatable than that negotiated under duress the previous fall, which guaranteed a traumatic dilution in their share values. Under the "bail-out deal", Ottawa and the banks proposed to invest $500 million each in the form of debentures which could be converted to common shares on a sliding scale, starting as low as $2.50. The result would be the possible creation of another 500 million new shares, which would have a terrible impact on present holdings.

Dome's prime concern was the need for breathing space—probably a full decade or more—to repay the principal on its debt. But that alone was not enough: the company also required an infusion of new money to keep it buoyant, although financial analysts had differing views on the amount required. At a minimum it was several hundred million dollars. For months Dome had been taking delicate confidential soundings from Canada's investment fraternity concerning its chances of raising the money it needed from individuals and the big institutional investors such as pension funds and insurance companies. Its objective was straightforward and along the lines that Jack Gallagher mused about shortly after his return from the climactic Toronto meeting in September 1982: if Dome could itself raise a substantial sum—and it could not even make the effort without the concurrence of its bankers—it was just possible it would remain master in its own house and avoid touching the one-billion-dollar rescue fund offered by the banks and the government.

If dilution could be avoided, Dome's own shareholders would have most to gain. Some had paid as much as $25.38 for their shares at the company's market high early in 1981, and their investment was worth not much more than $6.50 a share about two years later. They had no desire at all to see the stock fall further in value, which it certainly would if the "bail-out" terms were accepted. Yet, no one knew if shareholders would be willing to pour more money into the troubled company to help safeguard their investment.

If it came about that the federal government and the banks took over direction of the company's affairs through the appointment of their retainers to a majority on the board of directors, the alliance could prove to be an uneasy one. The banks, whose interest is essentially conservative *now*, would want Dome to operate in an unadventurous manner, taking as few risks as possible, husbanding its resources, and safeguarding the banks' money. That, however, is not at all what the federal government would want. Ottawa's interest is to see Dome continue its search for hydrocarbons in the high-risk frontier areas of Canada (which it does so well), delineating inventories of future wealth against the day when the country will need high-cost oil from our remote regions. The two interests are fundamentally at odds, and a divisive tug-of-war could well develop in a boardroom dominated by the nominees of politicians and bankers.

There is no neat conclusion to Dome's story. The star performer has left the stage, other actors have taken their positions, and newcomers stand fidgeting in the wings. Even the nature of the performance is ambiguous: at first it was high adventure, then for a while it seemed to be a tragedy. Now apparently it is a long-running drama with sudden new twists and countless sub-plots. The oil play continues.

Afterword

On December 1, 1983, a stocky, ruddy-faced Scot named John Howard Macdonald, the new boss of Dome Petroleum, squinted uncomfortably into television lights at a crowded press briefing at the Delta Bow Valley Inn in Calgary and sketched out his prescription for the patient's return to health.

He had just come from a meeting of 165 creditors and government officials where he had made proposals for a refinancing of Dome to set the company back on its way to good health. Although details of Dome's ideas occupied a thick book, the essential elements were unchanged from those that had been worked on for months by its former management: the continued sale of assets, an appeal for more time to repay the company's enormous debts, and a new equity offering.

The important new factor in the equation—perhaps the key element that could make the plan work—was Macdonald himself since he possessed that most perishable attribute, forfeited by both Jack Gallagher and Bill Richards in the turmoil of the HBOG debacle: credibility.

Macdonald is unmistakably a top-drawer company doctor brought in, at considerable expense, to perform surgery and administer nostrums to the ailing Dome. A plain-talking chartered accountant with thick accent, dry humor, and no-nonsense manner, at fifty-five John Howard Macdonald was within five years of retirement as the chief financial officer of the Royal Dutch/Shell Group of companies in Europe when he was offered the formidable job of nursing Dome back to health.

He had spent twenty-three years with the sprawling Shell empire and enjoyed a highly paid job with considerable prestige in the world's second-largest multinational oil company. He knows as much about oil industry financing as it is possible for any one individual to absorb.

Macdonald is precisely the type of person that Dome's creditors insisted the company should recruit to see it through its troubled times: an astute money man who has rubbed shoulders with bankers for decades, speaks the patois of international finance fluently, and is neither enamoured with blinkered visions of Arctic resource development nor tainted with the disastrous legacy of expansionism run wild.

Since few senior executives are entirely innocent of ego, it is not unfair to assume that Macdonald regards the formidable challenge of breathing life into Dome as a potential triumph with which to crown a career that had taken him high indeed up the executive ladder but just short of the top rung. At Dome he is clearly in command, albeit with limited manoeuvering ability, and he has the opportunity to establish a reputation as a corporate miracle worker, an outsider brought in to perform a job that several Canadian business executives were offered but none had the stomach to tackle.

The inducements were not inconsiderable either: an unusually handsome package of salary and benefits and, perhaps more importantly, the option to purchase three million Dome shares at their market value on the day he joined the company, October 1, 1983, when they traded at about five dollars. If Howard Macdonald, who holds the titles of chairman, chief executive officer, and chief financial officer, can find the magic formula to move Dome's share price up just one dollar during his tenure in Calgary, he will retire a very wealthy man.

The search for Jack Gallagher's successor as chief executive, conducted by a small committee of Dome Petroleum directors, was headed, initially, by Winnipeg businessman Frederick (Wick) Sellers. Sellers, however, for a time entertained vain ambitions of his own that he might land Gallagher's job, and so he soon excused himself from the search committee. Gallagher also wanted to be part of the group, but his participation wasn't encouraged by the others.

The talent hunters, finally headed by Toronto lawyer Fraser Fell, chairman of Dome Mines Ltd., cast a world-wide net and were willing to recruit a top-notch executive from outside the oil industry if necessary. It was not long before Macdonald emerged as a prime candidate.

Fell, together with Allen Lambert, the former chairman of the Toronto-Dominion Bank who had become a Dome director just months earlier, went to London in the spring of 1983 to see Macdonald. "We had several meetings with him. Certainly he was impressive right from the start," said Fell.

Macdonald did not decide hastily however to jump from the comfortable Shell ship. He wanted to make his own inquiries about Dome. He also was on the move a great deal, and so Fell and Lambert had to be patient before getting their man's answer. Macdonald's name and those of other candidates were presented to Dome's major bank creditors and the federal government for approval. Macdonald, who would need to apply formally for landed immigrant status before he could work in Canada, passed muster without difficulty.

The new chief executive acquitted himself well in his early meetings with Dome's creditors, and at his first press conference as head of Dome, he left an impression of considerable competence. His diagnosis surprised no one, however: Dome had great operating strengths and its problem was basically a financial one requiring a financial, not an operating, solution.

The company would concentrate its efforts on what he called its "core assets." It intended to make money from three sources: its oil and gas business in Alberta and neighboring provinces, its "outstandingly good" gas liquids business, and the profitable contracting work that the Canmar fleet was doing in the Beaufort. "These are the businesses which are going to generate the income, the cash, to resolve the financial problems over a period of time," he said.

Macdonald announced that Dome had just disposed of its U.S. properties, long on the auction block, to Texaco, and (perhaps more important symbolically than for the revenue it generated) it had finally sold its very fancy G-II executive jet, the plane in which Bill Richards made the trip to Stamford, Connecticut, to negotiate the first half of the fateful HBOG deal and an embarrassing reminder of the company's former high-flying days.

Macdonald said to no one's surprise that both Davie Shipyards in Quebec and the Cyprus Anvil mine in the Yukon were being offered for sale while acknowledging that the market was soft for money-losing mining companies and shipyards subsisting only on government largesse. Also still for sale was Dome's twenty-three per cent interest in Sovereign Oil & Gas, an operator in the North Sea. The company also hoped, said Macdonald, to sell ten million of its 30.5 million shares in Dome Mines and keep the balance.

By this time, Dome had already raised $700 million by selling off parts of its enormous empire, and it hoped to cut its debt by about another $600 million by further asset sales, which would still leave it about $5.6 billion in debt. Unable to finance its own exploration activities, it had farmed out much of its extensive acreage, both in the Arctic and in Western Canada, in July of 1983 to a group headed by Home Oil Company, Ltd. (and including Dome Canada, Ltd.) which planned to spend about $1.47 billion on Dome lands over the next three years.

The central problem facing Dome was its need to reschedule over $2.2 billion in current debt repayments. It was still exceptionally vulnerable to changes in the cost of money; a mere one per cent rise in interest rates meant another $50 million bill for the company. Macdonald had asked the banks to agree to stretch repayments over a ten-and-a-half- to twelve-year period. With soaring optimism for a near-bankrupt, Dome also hoped to raise a record $700 million in equity-related securities, of which its major shareholder, Dome Mines, would buy $200 million—providing it could slip the noose on an irksome $225 million loan guarantee it had given to the Toronto-Dominion Bank to cover a portion of the oil company's debt.

The financial fix also included a request to the banks to increase the company's $245 million operating line of credit by a further $200 million—for which consideration Dome brashly offered no security.

Until new arrangements were made, the traumatic agreement-in-principle (the so-called bail-out deal) signed in Toronto in September of 1982, but still not ratified by Dome, four major Canadian banks, or the Canadian government, would remain in place as a safety net.

Macdonald said that if everything went reasonably well, and the banks concurred quickly, the refinancing might be completed between April and June of 1984. As this afterword was being written, it was doubtful that such a timetable could be met; for example, Citibank of New York, which heads the consortium of international banks that financed the second half of the HBOG takeover, indicated unhappiness with Macdonald's initial proposals. Dome's finances remained in an appalling mess, and whether John Howard Macdonald could unscramble the omelet cooked by his predecessors was still very much an open question.

Just a few blocks away from the hotel where Macdonald was enduring unaccustomed scrutiny from the Canadian news media, Bill Richards, the man who had been groomed to be Dome's chief executive and who had made many of the major decisions for years before his fall from grace, was quietly starting a new career in a suite of offices in the Bow Valley complex. His cigar smoke didn't quite mask the smell of fresh paint, and furniture had been piled up to make way for carpet layers.

He had set up his own company, Richards Petroleum Resources, through which he planned to raise money and put together oil exploration deals. He was also involved in the promotion of a plan to build a covered sports stadium for Calgary. It would be privately financed although its proponents would take advantage of any government funding programs that happened to be available. "It just wouldn't be sensible not to," Richards confided without a trace of irony. His only residual connection with the company where he had spent twenty-six years was as a consultant on the controversial proposal to ship liquified natural gas from Alberta and British Columbia to Japan, a project he said was "near and dear to my heart." This grandiose plan was not yet embraced with quite the same warmth by Howard Macdonald, and its future was far from assured.

Having severed his formal connection with Dome, Richards was anxious to put distance between himself and the world he had so recently left. When he met his former subordinates socially, he would discourage them from talking about Dome by insisting, "I'm not paid to listen to that crap."

Reflecting on his turbulent career, he now insists he never had any burning aspiration to be Dome's chief executive, a surprising claim given the unremitting vigor with which he tackled the second spot and his love of rapid-fire decision making. He admits, however, he was vain enough to believe he was the best man for the job even after Dome's brush with bankruptcy and the famous bail-out deal of September 1983. "I thought maybe the board would see things my way and maybe they wouldn't. I must tell you I was realistic enough to know that when a company gets into trouble, the classic way of dealing with it is to clean out existing management and find someone else. That's right out of the textbook. I am not sure if I were in their shoes I would have done anything different."

Having maintained a facade of impregnable good humor in Dome's darkest hours, now that he has left the firm he will admit: "There were times when that job was pretty terrifying. It created an enormous stress level. You would wake up in the middle of the night in a cold sweat because the consequences [of bankruptcy] were so awesome; the impact on the whole Canadian economy, on the employees who would have lost their life savings, and the 60,000 suppliers, many of whom would have gone broke. You would have to be a fool or insane not to be gravely concerned about it—and I was." At the same time, says Richards, you could not have the president going around with shaking knees.

Richards' departure from Dome, long anticipated, was finally announced in mid-September, to take effect at the end of that month. He was replaced as president by John Beddome, the company's executive vice president, who had run the day-to-day operations for years. A group of Richards' closest associates within the company—senior executives and members of his takeover team—threw a surprise farewell party for him on Saturday, September 24, 19-- at his 1,600-acre ranch about forty miles outside Calgary. They had secretly arranged for an oil pump to be erected on his land with the inscription, "W.E. Richards Discovery Well # 1." But they hadn't found oil for him—the pump merely produced water!

The group of about twenty people stood around for a while in the late fall sunshine with drinks, posing for photographs before the oil pump and swapping stories about old times. Then they adjourned to Richards' rumpus room where they threw a lively roast for their departing boss. His gifts included a gunslinger's belt with the holster containing a telephone (to which he admits addiction) instead of a six-shooter. He was presented with cowboy chaps, and since he is forever losing them, he was given a new briefcase, which his staff firmly handcuffed to his wrist. It was a boisterous evening; at one point Richards was photographed with a brown paper bag over his head.

He also collected a large colored cartoon, a gift from his takeover team, which he now displays proudly in his new office. The logos of the various companies Dome acquired under Richards' aggressive tutelage— Siebens, Mesa, Kaiser, and the infamous HBOG—are all there. Richards is pictured, feet up on the desk, calling for his hit list. "See, he even put horseshit on my shoes!" he explains, delighted at the caricaturist's exacting attention to detail. The presence in Richards' office of that particular cartoon, a jocular reminder of his unrestrained acquisitiveness, furnishes an eloquent insight into his view of his past: the brazen boardroom pugilist is unrepentant, insisting that Dome was laid low by ill fortune as much as his monumental imprudence. He says he doesn't believe much in looking back anyway (an odd assertion for a reader of history, but understandable given the view!); he is really only interested in the future.

For Jack Gallagher, at sixty-seven, it was too late for a new career and a chance to rehabilitate a shopworn reputation; in any event, he had long since lost his zest for business. For a few months after he vacated the chief executive's office at Dome Petroleum, he retained the chairmanship and the chief executive's title at Dome Canada, its exploration arm, but he also relinquished these duties at the end of 1983 to Fred McNeil, the former chairman of the Bank of Montreal.

After a life of unremitting activity Gallagher actually took a vacation in California, for him an almost revolutionary change of life-style. Despite Dome's lately acquired notoriety, however, he discovered to his pleasure that he was still in demand as a public speaker, and he made use of his opportunities to proselytize his views on Canadian nationalism and particularly senate reform. Honors still came his way: he was made an officer of the Order of Canada, and the University of Manitoba, where he studied geology almost half a century before, conferred on him an honorary doctor of science degree; he had been given an honorary doctor of law degree by the University of Calgary some years before. He also appeared before the ponderously titled Royal Commission of the Economic Union and Development Prospects for Canada. To its chairman, former Liberal finance minister Donald Macdonald, an old acquaintance who had introduced the controversial "super-depletion allowance" in the "Dome budget" of March 1977, he argued that the federal government should establish royalty and tax holidays to encourage industry to make the massive investments needed to explore for and develop Canada's great hydrocarbon and mineral potentials in the remote areas of the country. For Macdonald—and for other listeners who had heard Gallagher's plea for frontier resource development many times—it must have seemed like *déjà vu*. The old promoter, credibility impaired but vision unclouded, was still in there pitching.

— Jim Lyon

BIBLIOGRAPHY

Much of the information in this book came from hundreds of personal interviews. Where appropriate, those interviewed have been named in the text; in many instances, however, knowledgeable people were only willing to discuss Dome given the promise of anonymity.

Scarcely a day has gone by during the past three or four years when the name "Dome" has not appeared in newspaper headlines. The author's reading list included *The Financial Post*, *The Financial Times*, *The Globe and Mail*, *The Calgary Herald*, *The Vancouver Sun*, *The Province* (Vancouver), *The Toronto Star*, *Mcclean's* magazine, *Canadian Business*, *Energy Magazine* (sadly, now defunct), *Oilweek*, *Saturday Night*, *Alberta Report*, *Fortune*, and *Business Week*. Two *Fortune* magazine articles were especially helpful for insights into the first part of Dome's takeover of the Hudson's Bay Oil and Gas Company, and Dome's relationship with its bankers. For those who enjoy the intrigue of takeovers, the November 1981 issue of *The American Lawyer* gives an exceptionally detailed and entertaining account of the ultimate takeover of Conoco by E. I. du Pont de Nemours Corporation, an action triggered by Dome's pursuit of HBOG. Numerous reports by financial analysts from Canadian and U. S. brokerage houses were also helpful.

The following books, articles, speeches, etc., were all enlightening:

Anderson, Allan. *Roughnecks and Wildcatters*. Toronto: Macmillan of Canada, 1981.

Berger, Mr. Justice Thomas R. *Northern Frontier, Northern Homeland: The Report of the Mackenzie Valley Pipeline Inquiry*. Volume 1. Toronto: James Lorimer, 1978.

Dacks, Gurston. *A Choice of Futures: Politics in the Canadian North.*Toronto: Methuen, 1981.

Dosman, Edgar J. *The National Interest: The Politics of Northern Development, 1969-75*. Canada in Transition Series. Toronto: McClelland and Steward, 1975.

Foster, Peter. *The Blue-eyed Sheiks: The Canadian Oil Establishment*. Toronto: Collins, 1979.

————. *The Sorcerer's Apprentices: Canada's Super-Bureaucrats and the Energy Mess*. Toronto: Collins, 1982.

Gallagher, J. P. "Crude Oil Self-Sufficiency for Canada by 1990?" Excerpts from an address to the Toronto Board of Trade, January 1981.

Gray, Earle. *The Great Canadian Oil Patch*. Toronto: Maclean-Hunter, 1970.

Harrison, G. R. "Implications to Canada of Arctic Marine Technology." Speech to the Canadian Shipping and Ship Repairing Association, Montreal, February 1981.

————. "The Arctic—A Sea of Opportunities." Speech to the Fourth National Marine Conference, Vancouver, November 1981.

Livingston, John. *Arctic Oil—The Destruction of the North?* Toronto: Canadian Broadcasting Corporation, 1981.

Pimlott, Douglas; Brown, Dougald; Sam, Kenneth. *Oil Under Ice: Offshore Drilling in the Canadian Arctic*. Canadian Arctic Resources Committee, Ottawa, 1976.

Richards, William. "Prospects for the Eighties." Speech on the listing of Dome Petroleum Ltd. stock on the London Stock Exchange, London, June 2, 1980.

Todd, M. B. "Development of Beaufort Sea Hydrocarbons—An Opportunity for Canadian Industry." Paper presented at Montebello, Quebec, 1981.

Annual Reports:
Dome Exploration (Western) Ltd., 1950-57
Dome Petroleum, 1958-82
Dome Canada, 1981-82
Hudson's Bay Oil and Gas Company, 1980
Conoco Inc., 1980

APOA Review. Arctic Petroleum Operators' Association.

Beaufort magazine (now defunct), published by Dome Petroleum Ltd., Esso Resources Canada Ltd., and Gulf Canada Resources Inc.

Dome Canada prospectus, 1981.

Dome Petroleum Forms 10-K and 10-Q filed with the Securities and Exchange Commission, Washington, D.C.

Dome Petroleum submission to the Special Committee of the Senate on the Northern Pipeline, 1982.

Environmental Impact Statement on Hydrocarbon Development in the Beaufort Sea—Mackenzie Delta Region. Dome Petroleum Ltd., Esso Resources Canada Ltd., and Gulf Canada Resources Inc., 1982.

The National Energy Program 1980. Energy, Mines and Resources Canada.

Plan of Arrangement. In effect a detailed prospectus concerning the merger of Dome and HBOG, mailed to shareholders of Hudson's Bay Oil and Gas Company Ltd. on or about December 18, 1981.

Sailing Directions: Arctic Canada. Department of Fisheries and Oceans, Ottawa.

INDEX

Alaska, 106–108, 114, 117, 142–143, 169

Alberta Report magazine, 203

Alcohol, bootlegging in the North, 157–158

Allen, Roger, 158, 177

Amazon, 32, 35

American University, Cairo, 34

Amoco Canada Ltd., 72

Andes, 34, 37, 51

Andruik, John, 188

Arctic
drilling, 62, 63, 100, 101, 102, 105, 108, 110, 112, 115, 116, 123, 134, 137, 138–139
fleet, planned sale, 192
land claims, 99
sovereignty, 62–63, 64–65
strategy, 61, 62, 101
supertankers, 119, 120, 123–124

Arctic Islands, 61, 65, 98–99, 101

Arctic Marine Locomotive, 60–61

Arctic Petroleum Operators' Association, 171

Arctic Pilot Project, 101

Arctic Waters Pollution Prevention Act, 64

Bailey, Ralph, 14–16

"Bail-out" deal, 29, 68, 186, 194–195, 197–198, 199–200, 201–202, 222, 226, 228, 235

Bandeen, Robert, 73, 89

Bank of Montreal, 12, 184, 188, 193, 201

Bank of Nova Scotia, 184, 188

Barclay, Ian, 21, 22

Beaufort Sea, 8, 10, 20, 52, 61–66, 98, 101, 102–104, 108, 111, 124, 128, 161, 169, 170, 180, 209
drilling acreage, 65, 101, 110
projected oil production, 118–122

Beaufort Sea Community Advisory Committee (BSCAC), 169, 170, 172

Beddome, John, 73, 87, 199, 224, 236

Bennett, Prime Minister R. B., 31

Bennett Jones, 76

Berger, Mr. Justice Thomas, 174

Berger Royal Commission, 119, 168

Bering Strait, 131, 144

Big Four Banks, 185, 186, 190,

199, 202, 204, 226
Bow Valley Industries, 19
Brazil, 145, 187, 217
British Columbia, 6, 49, 105,
    227
Bronfman, Charles, 16
Buchanan, Judd, 63, 169

Cairns, Ethel, 38, 41, 43–44,
    47, 52
Caissons, 131, 132, 136
Calgary Herald, 54, 55, 65, 160
Canada Development
    Corporation, 58
Canada Party, 55
Canadian Broadcasting
    Corporation (CBC), 52, 84,
    168, 172, 196
Canadian Hunter Explorations
    Ltd., 24
Canadian Imperial Bank of
    Commerce (CIBC) 11, 29,
    184, 185, 188, 194–199,
    200–201, 204
Canadian Investment
    Development Corp., 45
Canadian Marine Drilling Ltd.,
    58, 104, 108, 109, 139, 167,
    170, 224, 233
  dredging, 130–131
  earnings, 109
  fleet, 7, 98, 101–102, 103,
    105–106, 110–111, 137–
    139, 141–142, 143, 192
  pay rates, 147–148, 178
Canadian National Railway
    Co., 73–76
Canadian Ownership Account,
    201–202
Canadian Pacific Investments,
    73
Canadian Pacific Railway
    (CPR), 61, 73, 81, 117

Can-Dive Services Ltd., 149,
    153
Canmar. *See* Canadian Marine
    Drilling Ltd.
Canmar Carrier, 144
Canpar, 77
Chevron Standard Ltd., 42, 66
Chrétien, Jean, 62, 68, 103,
    190, 199, 200
Citibank, 23, 186, 189, 192,
    203, 225–226, 235
Cohen, Mickey, 191, 195, 200
Collins, Mary, 171, 172
Columbian Oil Company, 36
Columbia Gas Development
    Corporation, Delaware, 101
Columbia Gas Systems, 72
Committee for Original
    People's Entitlement
    (COPE), 170–172
Conoca Inc., 13–19, 22, 184,
    213
Consolidated Gas Supply
    Corporation, Pittsburgh,
    101
Cotterill, Ewan, 62
Cournoyea, Nellie, 172
Coutts, Jim, 59
Crew change, 135, 141, 164
Crowe, Marshall, 213
Cunningham, Clive, 145
Cyprus Anvil Mining
    Corporation, 214–216, 217,
    226, 234

Dallas, Texas, 12
Daniel, Bill, 225
Danielewicz, Ben, 128, 129, 163
Davie Shipbuilding Ltd., 53,
    140, 141, 209, 234
Debts, 183–186, 189–190,
    195–196, 205, 212–215,
    225, 231, 234

Degolyer and MacNaugton Inc., 116

Dempster Highway, 158

DEW line, 156

Divers, 148–154

Dome Canada, 8–11, 24, 67, 109, 186, 192–193, 195, 220, 223, 226

Dome Exploration (Western) Ltd., 39, 41

Dome International, 187

Dome Miners, Ltd., 40, 41, 47, 60, 195, 200, 205, 214, 227

Dome Shareholders' Protection Committee, 206

Dow Chemical of Canada Ltd., 24, 74

Drake Point, Arctic Islands, 101

Dredging. See Canadian Marine Drilling Ltd.,

Drill Ships
commitment to buy, 60
description, 137, 138
life on board, 141–143
purchase, 103–104

Drilling Authority, 62, 63, 64, 105–106, 109, 222
charges, 109
rigs, 6, 134, 136, 138

Drumheller, 42

Dry dock, 124, 127, 138, 144, 149

Dunkley, Charlie, 42, 43, 46, 98

Ecuador, 34–37

Edmonton, Alberta, 39, 52, 71, 72, 110, 124, 134, 155, 159, 164

Edmonton Journal, 160

Education in the North, 176, 177

Egypt, 33, 34, 187

Eisenhower, Harry, 86, 188

Ellesmere Island, 124, 129

Environmental Impact Statement, 119

Esso Resources Canada Ltd., 112, 114, 118, 119, 130, 163, 175

Explorer I, 173

Faro, Yukon, 214, 226

Fell, Fraser, 213, 233

Frazee, Rowland, 184, 200

Frontier Exploration Allowance. See "Super-depletion"

Gallagher, Jack
Canadian nationalism, 43–44, 51, 54, 58–59, 238
education, 30–33, 38, 238
fascination with North, 31, 98, 209
health, 29–30, 37–38, 48, 49–50
lifestyle, 17, 18, 37–38, 46–48, 51–52, 238
lobbying, 55–60, 108, 238
management style, 29, 41–42, 48–51
opposes HBOG takeover, 208–209, 210
personal fortune, 29, 38–39, 43, 46, 223–224
proposals for parliamentary reform, 54–55, 238
sensitivity to Northern issues, 171–172
"Smiling Jack", 30

Gallagher, Kathleen Marjorie, 48, 49

"Gallagher Amendment". See "Super-depletion"

"Gallagher special", 43

Galveston, Texas, 105

Gambling in the North, 158

Geological Survey of Canada, 31, 97, 98

Gillespie, Alastair, 57, 60, 63, 73

Global Marine Inc., 103, 109

Globe and Mail, 160

Graham, Terry, 127

Gray, Earle, 53

Gray, Herb, 5

Gray, Jim, 211

Great Slave Lake, 32

Greenhill, Robert F., 15

Gautemala, 34, 35, 56

Gulf Canada Resources Inc., 70, 98, 103, 109, 112, 114, 118, 119, 175

Gulfstream jet, 18, 84, 199, 234

Hans Island, N.W.T., 128, 129

Harrison, Gordon, 58, 61–63, 104–108, 113, 117, 139, 167–171, 224

Harrison, Russell, 184, 185, 194, 195, 197, 200–201, 203

Haskayne, Dick, 18

HBOG. *See* Hudson's Bay Oil and Gas Company

Hecla Gas Fields, 101

Heritage Savings Trust Fund, 53, 198

Herschel Island, N.W.T., 146

Hibernia, 46, 66, 139

Home Oil Company, Ltd., 234

Hopper, Wibert, 18

Houston, Texas, 12, 224

Hudson's Bay Company, 13, 20, 22, 77, 156

Hudson's Bay Oil and Gas Company (HBOG), 4, 12–16, 18–25, 68, 74, 81, 93, 183, 185–189, 190, 207–

210, 213–215, 216, 217

Humble Oil, 64, 107

Hunt brothers, 102, 109

Hunting and trapping, 161, 162, 166, 174, 178

Huskey, Susie, 167, 169, 170

Ice, 123–129, 133, 140, 146, 161

Icebreaking supertankers, 60, 63, 101, 118, 119, 123, 137–140, 179

Imperial Oil Ltd., 18, 37, 38, 41, 42, 64, 90, 97, 98, 106, 112

Indonesian oil fields, 74, 187

Inspector General of Banks, 185

Interest rates, 19, 183, 204, 207, 212, 213, 225

International Energy Agency, 58

Interprovincial Pipe Line Ltd., 72

Inuit, 63, 130, 133, 144, 155, 157, 159, 162, 163, 164

Inuvik, 10, 52, 127, 136, 142, 157, 158, 167, 176

Jacobson, Jimmy, 162, 163

Janson, Bill, 131, 132

Japan, 112, 127, 137, 159, 227

Japanese investors, 36, 83, 187, 193, 226

Japanese National Oil Company, 10

Jewell, Carl, 143, 144

Johnston, Dr. Don, 48, 87

Jones, Maclean, 199, 213

Kaiser Resources Ltd., 74, 85, 90, 210

Kawasaki Heavy Industries, 140

Kenny, Colin, 190
Kigoriak, 126, 136, 137, 145, 149, 152
King Edward Hotel, 22, 190, 199
Kugamallit Bay, Mackenzie Delta, 166, 174

Lalonde, Marc, 5, 9, 23, 67, 68, 190, 192, 194, 195, 197, 199
Lambert, Allen, 233
Land claims, native, 170
Land holdings, 76, 77, 101, 110, 222
Latimer, Radcliffe, 73, 74
Latin America, 34–37
Lauk, Gary, 195, 196
Lauzon, Que., 53, 140
Leduc, 38, 41, 42
LeMeur, Father Robert, 175, 176, 178, 179
Liberal Party, 5, 59, 190
Life magazine, 100
Liquefied natural gas, 101, 119, 208, 227
Liquefied petroleum gas (LPG), 216
Loeb, John, 41, 199
Loeb, Rhoades and Company, Inc., 41, 45, 47
London Stock Exchange, 69
Los Angeles, 102
Lougheed, Peter, 52, 197, 217

McCrea, Jim, 41, 205
Macdonald, Donald, 64, 238
Macdonald, John Howard, 231–234
MacEachen, Allan, 5, 23, 89, 192
McGrath, Wayne, 75–78
Mackenzie Delta, 10, 61, 97, 98, 112, 117, 118, 142, 155, 166, 175, 178
Mackenzie River, 133, 142, 155, 157
Mackenzie Valley Pipeline, 118, 167
Mackenzie Valley Pipeline Inquiry, 118, 173
McKinley Bay, N.W.T., 7, 115, 126, 127, 137, 143, 160, 162
Maclean's magazine, 55
Maier, Gerry, 18, 19, 213, 215
Manhattan, 64, 107
Manitoba, 4, 81, 97, 105
Marchand, Jean, 107
Massey-Ferguson, 85, 195
Melville Island, N.W.T., 100, 101
Mesa Petroleum, 77, 210
Métis, 155, 160, 164
Michel, Cliff, 41, 99
Mitsui Bank, 226
Mobil Oil, 57, 103, 105
Monteith, Oran, 106,
Montreal, Que., 4, 146, 155, 178
Montreal Trust Company, 22
"Moon pool", 137
Morgan Stanley and Company, 14
Morton, Bill, 41, 199
Motyka, Dan, 112
Mulholland, Bill, 183, 201

Nanaimo, Vancouver Island, 150
National Energy Board, 61, 213
National Energy Program, 5–9, 11, 14, 67, 68, 109, 118, 195, 210, 214, 217
Natives of the North, 155–166, 167–179

Natural gas liquids (NGLs), 69, 72–74, 81, 227
New York, 14, 37, 58, 99, 105, 188, 203
Noranda, 100
North Africa, 33
North Sea, 63, 105, 106, 117, 143, 145, 152, 170, 187
Northwest Passage, 60, 63, 106, 118, 139, 140
Northwest Utilities Ltd., 72
Nova Corporation, 14
Nuttall, John, 163

Ocean Ranger, 174
Oil
  Canadian self-sufficiency, 5, 61, 114, 120, 121, 207
  conservation, 5
  industry, Canadian ownership, 7, 200, 210
  industry, Canadianization of, 5, 67, 68
  production cost, 121
  transportation, 118, 139, 140
Operation Amoeba, 8
Operation Swampy, 4, 13
Organization of Petroleum Exporting Countries (OPEC), 58, 61, 120
Ottawa, as seat of Canadian government, 4–7, 9, 11, 23, 54, 59, 60, 61, 63, 64–68, 81, 100, 102, 103, 105, 108, 133, 167, 172, 192, 194, 195, 196, 197, 200, 201, 217, 219, 228, 229

Pallister, Ernie, 171
Panama, 34
Panarctic Oils, 61, 100
PanCanadian, 72
Panhandle Eastern Pipe Line

Company of Houston, 101
Payne, Bill, 52, 53
Peat Marwick Mitchell and Company, 194
Peru, 32
PetroCanada, 6, 9, 12, 18, 60, 67, 101, 106, 195, 200, 201
Petrofina, 201
Petroleum Club, Calgary, 43, 47
Phelps, Mike, 67, 190
Pitfield Mackay Ross Ltd., 8
Polar bears, 69, 133, 145, 155, 161, 162, 166, 179
Political contributions, 59, 60
Prince Rupert, B.C., 53, 227
Provo Gas Producers Ltd., 71
Provost Gas Field, 42, 71
Prudhoe Bay, Alaska, 63, 106, 114, 117, 118
Psychological testing, 77, 79
Public share offering, 45, 219

Redwater oil field, 42
Richards, Bill, 81–93, 235–237
  differences with Gallagher, 88, 89
  early legal career, 81
  lobbying, 23, 67
  management style, 83–85, 236, 237
  share holdings, 89
Robert LeMeur, icebreaking supply boat, 137
Roughnecks, 141, 148
Royal Bank, 183, 188, 193, 196, 199, 212
Rundquist, Darryl, 150–152

Sachs Harbor, N.W.T., 158, 172
Saskatchewan, 6, 71, 203
Saskatchewan River, 42
Savidant, Steve, 17, 18, 23, 92

Sellers, Frederick W., 223, 232
Semi-submersible drilling caisson (SSDC), 115, 137, 139
Shearman and Sterling, 188
Shell Oil Company, 32, 33, 35, 106, 225
Shetland Islands, 169
Shipboard Ice Alert and Monitoring system (SIAM), 125, 126
Shoyama, Tom, 58, 64
Sideways Looking Aperture Radar (SLAR), 125
Siebens Oil and Gas Ltd., 75, 76, 208
Sikorsky Sky-crane, 136
Sinai, irrigation of, 34
Social issues, Dome's awareness of, 171–173
Socio-economic relations, 167–170, 172, 173
Spain, 146, 151
Sproule, John, 100
Stamford, Connecticut, 14, 16, 18, 207
Standard Oil Company of New Jersey, 33–35, 37–40, 97
Steelman Gas Ltd., 71
Straddle plant, 72
Strong, Maurice, 44–46
Summers Harbour, N.W.T., 149
"Super-depletion", 7, 64, 66, 238
Supplier III, 127
Supplier IV, 127

Takeover fever, 74–76, 210
Tar sands, 6, 117, 217
Tarsiut artificial island, 112, 130–136, 137, 162, 163
Tax-sharing, 66, 217, 219

Texaco, 233
Texas Gas Transmission Company, 101
Texas Instruments, 69
Texasgulf, 14
Thomson, Kenneth, 20
Thomson, Richard, 183, 185
Titanic, 125, 140
Toronto Board of Trade, 117, 210
Toronto-Dominion Bank, 183, 185, 188, 192, 195, 219, 234
TransCanada PipeLines Ltd., 10, 24, 60, 73, 75, 186, 197, 225
Trudeau, Prime Minister Pierre Elliott, 5, 34, 58, 67 190, 192, 196, 197, 199, 217
Tuk Base, 125, 155–159, 162, 163, 170, 173
Tuktoyaktuk, N.W.T., 112, 117, 123, 130, 136, 142, 145 155–166, 167–179
Tuktoyaktuk Peninsula, 7, 137

Ubico, General Jorge, 35, 56
U.S. Navy, 35, 99, 111, 149
U.S. Securities and Exchange Commission, 189, 192
University of Manitoba, 30, 81, 238

Van der Wal, Hans, 130, 131, 133
Victoria, B.C., 48, 87, 196

Wagner, Ardith, 44, 47
Washington, D.C., 108, 192
Watson, George, 199, 225
Well blow-out, 108, 133, 179

Winnipeg, 6, 28, 30, 47, 81, 199, 223

Wolcott, Don, 46, 69, 71

World Meteorological Organization, 125

Yellowknife, N.W.T., 32, 163, 166

Yukon, 142, 214, 215, 226

THE PENNY PINCHER'S WINE GUIDE

Lucy Waverman

For anyone who has tried to purchase just the right wine for a special occasion and been overwhelmed by the vast array of sparkling bottles—this is the guide that will make the wine selection crystal clear! Here are evaluations of over 500 wines, all available in Canada for under $8.00. Included are the "best" wine buys, a special section on Canadian wines and vineyards, and menus with recipes for all occasions.

Lucy Waverman, who holds an Advanced Certificate from Cordon Bleu, is a recognized authority on both food and wine. She runs her own cooking school in Toronto and has included some of her most outstanding recipes to accompany the perfect wines

An AVON Trade Paperback   85027-3/$5.95

STEPFAMILIES
MAKING THEM WORK

ERNA PARIS

Erna Paris, award-winning Canadian author, and a mother of a stepfamily, has written this comprehensive new volume filled with effective advice for coping with the unexpected problems of a new marriage with children. Based on actual interviews with stepparents and children, Paris provides an analysis of the successes and failures found in these real stories, offering practical answers to the questions raised by families encountering difficulties in the remarriage/stepparenting situation.

Remarriage with a stepfamily is a unique situation with a peculiar set of problems. In STEPFAMILIES: *Making Them Work*, Erna Paris addresses such complex issues as:

- When both partners bring children from a prior marriage into the remarriage, will they be able to live happily under one roof?

- Are there ways to avoid jealousies among the stepchildren?

- Can the children handle the disturbing conflict of sexual attraction within the stepfamily?

- Will your ex-spouse's involvement with your children cause friction in your new marriage?

- Do you come into the marriage as an intruder in the eyes of your spouse's children?

- And much more information to arm the stepparent with enough knowledge to enter this high-stakes arena and reap the great rewards possible through love and dedication.

An Avon Paperback   **86405-3/$5.95**
